THIRD EDITION

Electronic Components and Technology

D0206872

TUTORIAL GUIDES IN ELECTRONIC ENGINEERING

Series editors

Professor G. G. Bloodworth, *University of York*
Professor A. P. Dorey, *University of Lancaster*
Professor J. K. Fidler, *University of Northumbria*

This series is aimed at first- and second-year undergraduate courses. Each text is complete in itself, although linked with others in the series. Where possible, the trend toward a "systems" approach is acknowledged, but classical fundamental areas of study have not been excluded. Worked examples feature prominently and indicate, where appropriate, a number of approaches to the same problem.

A format providing marginal notes has been adopted to allow the authors to include ideas and material to support the main text. These notes include references to standard mainstream texts and commentary on the applicability of solution methods, aimed particularly at covering points normally found difficult. Graded problems are provided at the end of each chapter, with answers at the end of the book.

THIRD EDITION

Electronic Components and Technology

Stephen Sangwine

CRC Press
Taylor & Francis Group
Boca Raton   London   New York

CRC Press is an imprint of the
Taylor & Francis Group, an informa business

CRC Press
Taylor & Francis Group
6000 Broken Sound Parkway NW, Suite 300
Boca Raton, FL 33487-2742

© 2007 by Taylor & Francis Group, LLC
CRC Press is an imprint of Taylor & Francis Group, an Informa business

No claim to original U.S. Government works
Printed in the United States of America on acid-free paper
10 9 8 7 6 5 4 3 2 1

International Standard Book Number-10: 0-8493-7497-9 (Softcover)
International Standard Book Number-13: 978-0-8493-7497-5 (Softcover)

Library of Congress Cataloging-in-Publication Data

Sangwine, Stephen.
    Electronic components and technology / Stephen Sangwine. -- 3rd ed.
        p. cm.
    "A CRC title."
    Includes bibliographical references and index.
    ISBN-13: 978-0-8493-7497-5 (alk. paper)
    1. Electronic apparatus and appliances. 2. Microelectronics. I. Title.

TK7870.S275 2006
621.381--dc22   2006029159

Visit the Taylor & Francis Web site at
http://www.taylorandfrancis.com

and the CRC Press Web site at
http://www.crcpress.com

Contents

Preface to the Second Edition   vii

Preface   ix

Acknowledgments   xi

Author   xiii

1   Introduction   1

2   Interconnection technology   5
- Jointing   6
- Discrete wiring   11
- Cables   12
- Connectors   15
- Printed circuits   18
- Printed circuit assembly   24
- Rework and repair   25
- Case study: A temperature controller   26
- Summary   29
- Problems   30

3   Integrated circuits   31
- Review of semiconductor theory   33
- Integrated-circuit fabrication   35
- Semiconductor packaging   43
- Handling of semiconductor devices   45
- Custom integrated circuits   46
- Summary   50
- Problems   51

4   Power sources and power supplies   53
- Energy sources   54
- Batteries   55
- Power supplies   60
- Summary   74
- Problems   74

5   Passive electronic components   77
- Passive component characteristics   77
- Resistors   83
- Capacitors   87
- Inductors   91
- Summary   92
- Problems   93

6 Instruments and measurement 95
Quantities to be measured 96
Voltage and current measurement 98
Frequency and time measurement 101
Waveforms — The oscilloscope 104
Summary 109
Problems 110

7 Heat management 111
Heat transfer 112
Thermal resistance 113
Heat sinking 114
Forced cooling 119
Advanced heat-removal techniques 120
Summary 121
Problems 122

8 Parasitic electrical and electromagnetic effects 123
Parasitic circuit elements 123
Distributed-parameter circuits 128
Electromagnetic interference 133
Applications studies 142
Summary 149
Problems 150

9 Reliability and maintainability 153
Failure 154
The "bathtub" curve 155
Measures of reliability and maintainability 156
High-reliability systems 162
Maintenance 165
Summary 168
Problems 168

10 Environmental factors and testing 171
Environmental factors 171
Type testing 178
Electronic production testing 180
Summary 184
Problems 185

11 Safety 187
Electric shock 188
Other safety hazards 194
Design for safety 195
Summary 198

References 201

Answers to problems 203

Index 205

Preface to the Second Edition

This book is intended to support Engineering Applications studies in electronic engineering and related subjects such as computer engineering and communications engineering at first- and second-year undergraduate level. **Engineering Applications**, abbreviated as **EA**, is a term first used in the report of the Finniston inquiry into the future of engineering in the United Kingdom. Finniston used the terms **EA1** and **EA2** to refer to the first and second elements of a four-stage training in Engineering Applications, to be taken as part of a first-degree course in engineering. Later, the Engineering Council, established as a result of the Finniston report, expressed the concept of EA1 as

> An introduction to good engineering practice and the properties, behaviour, fabrication and use of relevant materials, systems and components.

and EA2 as

> Application of scientific and engineering principles to the solution of practical problems of engineering systems and processes.

These quotations are taken from *Standards and routes to registration* (second edition), otherwise known as SARTOR, published by the Engineering Council in January 1990. They are reproduced here with the permission of the Engineering Council, United Kingdom.

Although EA studies should be integrated into the fabric of a degree course, there is a need to draw out elements of practice to provide emphasis. It is intended that this book should be used as a source, complementing other texts, for such studies.

In the context of electronics, product design is an activity that begins, by and large, with **components** rather than **materials**. This is not to say that a study of materials is not relevant as a part of EA1, but as there are many existing texts covering the subject, **materials** has been excluded from this book in favour of more coverage of components.

The book begins with an introduction to electronic interconnection technology including wiring, connectors, soldering and other jointing techniques, and printed circuits. Chapter 3 is devoted to the very important technology of integrated circuits, concentrating on their fabrication, packaging, and handling. **Components** is taken to include power supplies, as in many applications a power supply unit is bought-in as a subsystem. The main characteristics of power supplies and batteries are covered in Chapter 4. Passive electronic components are introduced in Chapter 5, and with them the book begins to include a major theme developed in Chapters 7 and 8: the **parasitic effect**. This includes the nonideal properties of passive components introduced in Chapter 5, heat and its management in Chapter 7, and parasitic electromagnetic effects in Chapter 8. EA1 is essentially practically oriented and will include laboratory-based work, including the use of tools and instruments. A new chapter has been added to the second edition to add to the utility of the book in supporting EA1 studies and laboratory activities. Thus, Chapter 6 introduces the instruments and measurements used in electronics and related subjects. Chapter 9 reviews good engineering practice in relation to reliability and maintainability, two important aspects of design which, unfortunately, are often overlooked by electronic circuit designers. Chapter 10 introduces environmental influences on electronic products and the subject of testing both for environmental endurance and in production. The final chapter in the book introduces safety.

The book assumes that the reader has taken the first one or two terms of a degree course, although some of the earlier material could be studied sooner. Extensive cross-references to more specialized texts have been given in the marginal notes and the bibliography, including, where appropriate, references to other texts in the Tutorial Guides Series. These have been updated for the second edition to include later books in the series where appropriate, the latest editions of technical standards, new editions of books previously listed, and some additional books published since 1986. Other major revisions in the second edition include updated information in Chapter 3 to take into account changes in IC technology since 1986, changes to the final chapter to take into account new legislation, and some new illustrations.

Part of the aim of this book is to inform the reader about components, technology, and applications, but it is also intended to create an awareness of the problems of electronic engineering in practice. I hope that the readers of this book will be encouraged to tackle these problems and go on to become competent and professional electronics engineers.

Safety Note

The material in Chapter 11 is of course at an **introductory level only**, and readers are cautioned that professional competence in safe electrical design cannot be achieved merely by studying the contents of this chapter.

Preface

Since the publication of the second edition of this book in 1994, some very significant changes have occurred in technology, particularly the further miniaturization of electronic products, and the steady and quite dramatic increases in the speed of computers. In electronics, surface-mount technology has become almost universal and vastly better batteries have been developed for laptop computers and portable phones. Technically, however, electronics has not changed in revolutionary ways. When revising the book for this third edition, it was a surprise to discover that components and technologies featured in the second edition were still commercially available. Nevertheless, the revisions needed after 12 years were extensive, but they did not require the text to be restructured. Examples of the areas that needed updating are: the introduction of lead-free solders and digital oscilloscopes, and new types of batteries. The bibliography has been brought up to date, and all references to technical standards, European Union directives, and the like have been checked and, where necessary, updated.

The book has now been in print for 19 years. The previous editions were published in the United Kingdom and largely written for a British audience. In revising the book for this third edition, the opportunity has been taken to make the book more usable elsewhere in the English-speaking world, by small changes in terminology and vocabulary, and by reference to international standards, rather than British Standards, where applicable.

This new edition was prepared electronically, which should make it easier to update at reprinting if the publishers wish to do so. Therefore, please contact me with any corrections or suggested amendments at S.Sangwine@IEEE.org.

Stephen J. Sangwine
Colchester, United Kingdom

Acknowledgments

The first edition of this book developed from a lecture course that I first presented in 1985 as a part of new EA material introduced into engineering courses at the University of Reading. I would like to thank my former colleague Peter Atkinson for his early suggestion that a book could be written and for his help and encouragement while I wrote the book and subsequently. I would also like to thank S. C. Dunn, former chief scientist at British Aerospace, who made many helpful suggestions at an early stage, including the theme of parasitic effects. The following people contributed advice, criticism, or technical background during the preparation of the first edition, and I wish to acknowledge their help: Susan Partridge of General Electric Company (GEC) Hurst Research Centre for reading the first draft of Chapter 3; Alistair Sharp of Eurotherm Ltd. for help with the case study in Chapter 2; John Barron of Tectonic Products, Wokingham, for help with the subject of printed circuits; John Terry of the Health and Safety Executive and Ken Clark, deputy director at Baseefa, for both help and criticism of Chapter 10 (now Chapter 11); Dr. George Bandurek at Mars Electronics at Winnersh for commenting on Chapter 8 (now Chapter 9); and Martin Thurlow of the Electromagnetic Engineering Group, British Aerospace, and David Hunter of the Army Weapons Division, British Aerospace, for assisting with illustrations and background. I am also grateful to Carole Hankins for typing the manuscript and to my wife, Elizabeth Shirley, for her support over many months of writing and revising. Professor A. P. Dorey was consultant editor for this book and made many helpful comments on drafts of the manuscript.

During revision for the second edition, the following people contributed advice and I wish to acknowledge their help: Trevor Clarkson of King's College for suggesting the addition of measurement; John Whitehouse for suggestions in Chapter 8; and John Terry (again) for reading the former Chapter 10 and suggesting numerous updates, which I have incorporated into Chapter 11. I would also like to thank Charles Preston for helping with the illustrations, particularly for Chapter 6. Professor A. P. Dorey was, once again, consultant editor and made helpful comments on drafts of the new material.

During revision for the third edition, the following people assisted with illustrations and I wish to acknowledge their help: Mark Johnson of Megger Limited; Dr. Ursula Kattner of National Institute of Standards and Technology (NIST) for help with lead-free solders; Stephen Head and Dave Bremner of Eurotherm Ltd.; Marcus Brain of Technisher Überwachungsverein (TÜV) Product Services; Steve Smith of Schaffner Limited; R. J. Sullivan of Aavid Thermalloy; Jeff Weir, Naomi Mitchell, and Cole Reif of National Semiconductor; Paul Bennett of Bulgin Components; Greg Macdonald of Amphenol Canada and Gilles Dupre of Amphenol-Socapex, France; Natasha Moore of BEAB–ASTA (the British Electrotechnical Approvals Board and Association of Short-Circuit Test Authorities); Yuko Takahashi of Fujitsu Limited; and Gary Silcott of Texas Instruments.

Several companies and organizations have supplied illustrations or granted permission for me to use their copyright material, and they are acknowledged in the text. Permission to reproduce extracts from BS-EN 61340-5-1: 2001 is granted by the British Standards Institution (BSI). British Standards can be obtained from BSI Customer Service, 389 Chiswick High Road, London W4 4AL, United Kingdom; Telephone: +44 (0)20 8996 9001. The author thanks the International Electrotechnical Commission (IEC) for permission to reproduce information from its *International Technical Specification IEC 60479-1*, Fourth Edition (2005) and from its International Standard IEC 61340-5-1, First Edition (1998). All such extracts are copyright of IEC, Geneva, Switzerland. All rights reserved. Further information on the IEC is available from www. iec.ch. IEC has no responsibility for the placement and context in which the extracts and contents are reproduced by the author, and IEC is not in any way responsible for the other content or accuracy therein.

Author

Stephen J. Sangwine was born in London, United Kingdom in 1956.

He has a B.Sc. degree in electronic engineering from the University of Southampton, United Kingdom (1979), and a Ph.D. degree from the University of Reading, United Kingdom (1991). He is currently a senior lecturer with the Department of Electronic Systems Engineering at the University of Essex, Colchester, United Kingdom. From 1985 to 2000, he was a lecturer with the Department of Engineering at the University of Reading, where he wrote the first and second editions of *Electronic Components and Technology*. From 1979 to 1984, he worked in the civilian nuclear power industry at the United Kingdom Atomic Energy Authority's Harwell Laboratory, designing radiological monitoring instruments, including one of the very earliest applications of complementary metal-oxide semiconductor (CMOS) microprocessors to a pocket-sized instrument.

As well as authoring *Electronic Components and Technology*, Dr. Sangwine coedited *The Colour Image Processing Handbook* (Chapman & Hall, 1998), and he has also authored or coauthored more than 70 papers, the majority in the field of image processing.

His principal research interest is in linear vector filtering and transforms of vector signals and images, especially using hypercomplex algebras, on which he collaborates with researchers in the United States and France. In 2005, he was a *chercheur invité* (visiting researcher), Centre National de la Recherche Scientifique (CNRS), at the Laboratoire des Images et des Signaux, Grenoble, France, for 7 months, with financial support from the Royal Academy of Engineering, United Kingdom.

Dr. Sangwine has been a senior member of the Institute of Electrical and Electronics Engineers (IEEE) since 1990.

Introduction   **1**

Modern electronic engineering products are found in a wide range of applications environments from the floor of the deep ocean (submarine cable repeaters) to geostationary orbit (microwave transceivers on board communications satellites), from the factory floor (industrial process controllers and numerically controlled machine tools) to the office (computers and printers). They can be found in the home (audio and video systems and microwave ovens), in schools (computers and pocket calculators), in hospitals (computerized tomography [CT] and magnetic resonance imaging [MRI] scanners, bedside monitors), inside the human body (heart pacemakers), and inside road vehicles (electronic ignition and engine management, antilock braking). Electronic products can also be found in the pocket (portable phones, personal audio, and video players). These products may be mass produced by the million, or they may be one-off special systems. They may be intended to last for decades, or they may be designed deliberately for a fairly short life. They should all be fit for their intended purpose and be of significant use to their users.

All electronic products depend on the physical and electrical properties of insulating, conducting, and especially semiconducting **materials**, but by and large, the designer of an electronic product works with **components** and **technologies**, such as integrated circuit (IC) technology, rather than with basic materials. A critical aspect of product design is the interconnection of components, and for this reason this book starts with a chapter covering the technology of interconnection. The technology of interconnecting electronic components, circuits, and subsystems was, until the publication of this book, often neglected in electronic engineering texts at degree level. It is true that the detailed layout of a printed circuit board (PCB) is not a task likely to be undertaken by a graduate engineer unless the PCB is to carry high-frequency or high-speed circuitry. Nevertheless, a PCB has electrical properties and its design, together with the choice of components to go on it, can have a significant effect on the performance, the cost of production, the production yield, and the reliability and maintainability of the assembled board, and quite likely the product of which it is a part. Jointing techniques, especially soldering, are of tremendous importance in electronic engineering, and solder is an engineering material that should be specified as carefully as a mechanical engineer specifies structural steel: what **type** of solder is best suited to a particular application? In many cases, just "solder" will not do.

The third chapter deals with IC technology. Only a few engineers are involved in high-volume IC design, but a more significant number design or use semicustom ICs. Consequently, the treatment in this book is not for the IC specialist: it is aimed at the much larger group of electronics engineers who will be using ICs or designing a gate-array or standard-cell IC of their own.

Competent electronics engineers need a good understanding of the components and subsystems from which their designs will be constructed and the instruments needed to test and characterize prototypes. They must be aware of not only the ideal behaviour of components, subsystems, and instruments, but also their performance limitations. The next three chapters, therefore, cover power sources and power supplies (an important class of electronic subsystem), passive electronic components, and instruments and measurement. To understand the performance limitations of components, an engineer must appreciate how the components are fabricated. To understand the performance limitations of power supplies and instruments, an engineer must appreciate the principles on which they operate. Chapters 5 and 6 cover these topics as well as provide factual information for reference.

The study of electronic components introduces the third major theme of this book: the **parasitic** effect. Real electronic components and circuits, as opposed to **ideal** ones, possess parasitic properties that are incidental to their intended properties. A wire-wound resistor, for example, is also inductive and has an impedance that varies with frequency. **Heat** is produced in significant quantity in some electronic systems, and positive design measures often have to be taken to remove it. Electromagnetic energy can radiate from electronic circuits and couple into other circuits, causing faulty operation. A chapter has been devoted to this and other parasitic electromagnetic effects. This book does not attempt to cover all possible parasitic effects: to do so would be impossible even in a much larger book and would serve little useful purpose. Electronics engineers must learn to expect parasitic effects and try to take them into account when designing electronic products.

So far this introduction has dealt with matters that affect design and performance in ways that are important at the beginning of the life of a product. Without an understanding of components, technology, and parasitic effects, the design engineer will not be able to design good products that meet the required level of performance at the required cost. Many electronic products, however, will have a life that lasts far longer than the designer's interest in the design. It is during the operating life of a product that long-term effects become important. Components and materials **age**: they deteriorate physically and chemically, and ultimately they **fail**. The study of these problems and of the prediction of product life is known as **reliability**. Not surprisingly, the reliability of a product can be influenced by its design, for better or for worse, and if a product is capable of being repaired, the ease and expense with which it can be restored to working order can also be affected by decisions taken at the design stage. Reliability can also be influenced by a product's operating environment. Did the designers consider the effects of temperature, humidity, corrosion, and dust? Is there some unknown environmental factor that will doom their product to early failure? As with parasitic effects, after introducing some of the many environmental hazards to electronic equipment, this book leaves the readers to consider what the problems of their products' environment might be.

Lastly, this introduction has dealt with the electronic product itself: will it work and continue to work for long enough? Will it succumb to environmental stress? Engineers must also look at their designs from another viewpoint: will they do anyone, or the environment, any harm? All design engineers, including those working in electronic engineering, have a professional duty to consider safety when designing products, and in many countries a statutory (that is, legal) duty also. The final chapter introduces the subject of safety in electronic engineering.

Interconnection technology

2

Objectives

- ☐ To emphasize the importance of interconnection in electronic product design.
- ☐ To discuss jointing technology, especially soldering and solderless wire-wrapping.
- ☐ To outline the main types of discrete wiring and cabling.
- ☐ To describe the technology of printed circuits.
- ☐ To give an introduction to rework techniques.
- ☐ To present a short case study illustrating the importance of interconnection in industrial product design.

All except the smallest of electronic systems are built up from subsystems or subassemblies that are in turn built from electronic components such as resistors, capacitors, transistors, integrated circuits (ICs), displays, and switches. A desktop personal computer, for example, is likely to be built from a power supply subsystem, a main circuit board, and a number of peripheral subsystems such as a CD/DVD drive and plug-in memory modules. Small self-contained electronic products such as pocket calculators and portable phones are often built directly from components with no identifiable subsystems.

From the lowest level of component up to the system level, the constituent parts of an electronic system have to be interconnected electrically. The lowest level in the hierarchy of interconnection is the electrical **joint**. From the very earliest days of electronics, long before the invention of the transistor and integrated circuit, soldering has been an important technique for making electrical joints. Hand soldering is still used in prototype work, repair work, and, to a much lesser extent, production. Not all electrical joints in an electronic system need to be soldered: the technology of solderless wire-wrapping is well established in digital electronics, for both prototype and production wiring, and joints can also be made by insulation displacement, welding, or crimping. Components and subsystems are interconnected by wiring that can be in the form of either discrete wires and cables or printed circuits. A short case study at the end of this chapter illustrates the trend in electronic engineering over the last 20 years towards printed circuit interconnection wherever possible, avoiding discrete wiring because of the high cost of assembling and inspecting individual wires.

The printed circuit board, or PCB, is tremendously important in almost all applications areas of modern electronics. Not only does it provide a cheaply mass-produced means of interconnecting hundreds or thousands of individual components, but it also provides a mechanical mounting for the components.

In the very early days of electronics when thermionic (vacuum tube) valves were used, interconnection with discrete wiring was normal — but this was soon superceded with the advent of transistors by the introduction of printed circuit boards.

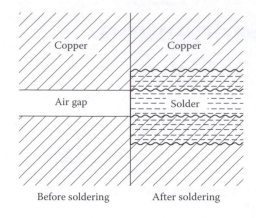

Figure 2.1   Diagrammatic representation of a soldered joint (not to scale).

Jointing

Several techniques are used in electronic engineering for making electrical joints. The most important of these is soldering, which is used mainly for attaching and jointing components to PCBs, but it is also widely used for jointing in cable connectors. Another important technology is solderless wire-wrapping, which finds application in prototype wiring for logic circuits and in production wiring of backplanes interconnecting PCBs. Welding is used in some specialized electronic applications and is a very important jointing technique in integrated circuit manufacture. Finally, in applications where soldering or welding cannot be used, mechanical crimping can make sound electrical joints.

Soldering

Solder is a low-melting-point alloy of tin and other metals, used for making electrical and mechanical joints between metals. Soldering does not melt the surfaces of the metals being joined, but adheres by dissolving into the solid surface. Figure 2.1 shows an idealized cross-section of a soldered joint illustrating this point. Soldered joints can be made by hand using an electrically heated soldering iron or by a mass-soldering process in which all the joints on a PCB are made in one automated operation. Both techniques are important, and they are described in detail below and in a later section of this chapter.

In the twentieth century, solders used in electronics were almost always alloys of tin and lead. From July 2006, the European Union (EU) has required lead (and many other toxic substances) to be eliminated from electronic and other products, and therefore lead-free solders are now used in place of tin–lead solders. Because of the historical importance of tin–lead solder in electronics, we start with its properties before progressing to the characteristics of lead-free solders.

Commercial tin–lead solders were available with several different proportions of tin to lead and with traces of other metals to enhance their properties. Tin and lead are soft metals with melting points of 232°C and 327°C respectively. Alloys of these metals generally start to melt at a temperature of 183°C, which is lower than the melting temperature

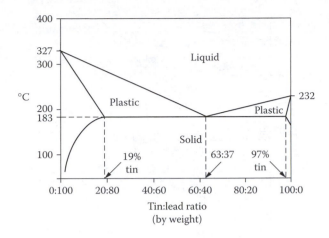

Figure 2.2　Phase diagram for tin–lead solder alloys (simplified).

of either pure metal. Figure 2.2 shows a simplified phase diagram for tin–lead alloys. The ratio (by weight) of tin to lead is plotted horizontally with 100% lead on the left and 100% tin on the right. The vertical scale represents temperature. In the top region of the diagram, above the line extending from the melting point of lead at 327°C across to the melting point of tin at 232°C via the point at a tin:lead ratio of 63:37 and a temperature of 183°C, the alloys are liquid. The bottom region of the diagram represents the ranges of temperature and tin:lead ratio over which the alloys are solid. The two triangular regions represent temperatures and compositions where the alloys are in a plastic state consisting partly of solid and partly of liquid. The alloy with a tin:lead ratio of 63:37 is the only one that changes sharply from solid to liquid at a single temperature. This alloy is known as a **eutectic alloy**. It is fully liquid at the lowest possible temperature for a tin–lead alloy.

For general electronic jointing, a 60:40 solder was used, which is fully molten at about 188°C. 40:60 solder, which is fully molten at about 234°C, was also readily available for applications where a higher melting point was needed.

The most common lead-free solder for use in electronics is an alloy of tin (Sn), silver (Ag), and copper (Cu) in proportions of about 95 to 96% tin, 3 to 4% silver, and 0.5 to 1% copper. Alloys of this composition have a melting point of around 215 to 218°C, which is somewhat higher than the melting point of a 60:40 tin:lead solder. Figure 2.3 shows part of the phase diagram for the tin–silver–copper alloys (the full phase diagram takes the form of an equilateral triangle, but since a practical alloy for electronic soldering has 95 to 96% tin, only one corner of the full diagram is shown). Temperature is shown by contour lines, since to plot them would require a third dimension, out of the page. The higher melting point of lead-free solders means that more care has to be taken to control soldering processes, since the higher temperatures could more easily cause thermal damage to electronic components.

Four requirements must be met if a good soldered joint is to be made, and an understanding of these is essential in order to develop skill at hand soldering. First, the surfaces to be joined must be solderable. Not

Phase diagrams in general and the tin–lead phase diagram in particular are discussed by Anderson et al. (1990).

The proportion of tin (by weight) is conventionally stated first.

PCBs sometimes incorporate a solderability test pad in an unused corner of the board, so that the solderability of the board or a batch of boards can be checked before component assembly.

7

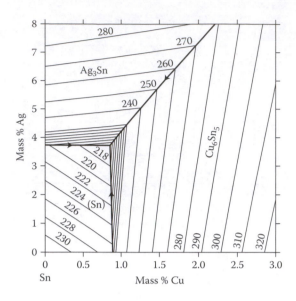

Figure 2.3 Partial phase diagram for tin–silver–copper lead-free solder alloys. (Courtesy of Dr. Ursula Kattner, National Institute of Standards and Technology [NIST], United States.)

Mercury is toxic. Do not try this experiment.

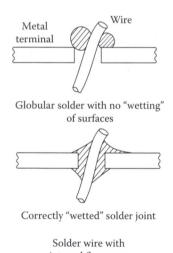

all metals are readily soldered without special techniques. Aluminium, for example, is an extremely reactive metal that rapidly oxidizes on exposure to air to form a passivating oxide layer that prevents solder from alloying with the underlying metal. It is possible to demonstrate the reactive nature of aluminium by removing the oxide layer with a little mercury rubbed onto the aluminium surface. The freshly exposed surface reacts rapidly with moisture in the air, and the metal becomes hot. Gold is very easily soldered because the metal does not oxidize. Copper and brass oxidize readily, but the oxide layer can be removed easily, so that these metals are solderable. The second requirement for a good soldered joint is cleanliness: the surfaces to be joined must be free of grease, dust, corrosion products, and excessively thick layers of oxide. In most electronics applications, the surfaces to be soldered will have been plated with gold or a solder alloy during manufacture in order to provide a readily solderable surface. Coating a metal with solder is known as **tinning**, and protects the metal from oxidation. Cleaning is not usually needed, therefore, unless a component or PCB has been stored for a long time or becomes contaminated with grease or dirt. The third requirement is that any layer of oxide on the surfaces must be removed during soldering and prevented from regrowing until the molten solder has alloyed with or "wetted" the surface. This is achieved by a **flux** that chemically removes the oxide layer and reduces surface tension, allowing molten solder to flow easily over the surfaces to be joined.

For hand soldering, solder wire with integral cores of flux is used. The fluxes commonly used for electronic work are rosin based and chemically mild, leaving a noncorrosive residue. More powerful fluxes based on acids may be needed on less solderable metals, but must be thoroughly cleaned off afterwards because they leave a corrosive residue. The final requirement for a good soldered joint is heat. Both surfaces to be joined must be heated above the solidification temperature of the solder, otherwise the solder will chill on contact and will fail to flow evenly and alloy with the surfaces. When soldering by hand, heat transfer from the soldering iron to the joint is improved if there is a little molten solder on the tip of the soldering iron. The joint should be heated with the iron, and the solder wire applied to the joint (not the iron). The iron should not be removed until the solder has flowed into the joint. It is very important that the joint is not disturbed until the solder has fully solidified, otherwise a high-resistance (dry) joint will result from mechanical discontinuities in the solder.

Wire-wrap jointing

Soldering is a good technique for mass jointing on PCBs, but it has several disadvantages for discrete wiring joints. An alternative jointing technique exists for logic circuits and low-frequency applications known as solderless wire-wrap. Special wire and terminal pins are used for wire-wrap jointing, and special hand-operated or electrically powered tools are required. Various types of IC sockets and connectors, including PCB edge connectors, are made with wire-wrap terminal pins. Wire-wrap interconnection is used for prototype and production wiring of logic boards and for production wiring of backplanes for interconnecting a rack of PCB subunits. It is a faster technique than soldering, requires no heat, produces no fumes, and can easily interconnect terminals spaced as little as 2.5 mm apart. Figure 2.4 illustrates a typical joint. The terminal pin is typically 0.6 to 0.7 mm square and 15 to 20 mm long. The wire is solid and about 0.25 mm in diameter. The joint consists of about seven turns of bare wire and one to two turns of insulated wire, wound tightly around the terminal pin. At each corner of the terminal pin, the wire and the pin cut into each other, making a metal-to-metal connection, which improves with age through diffusion of the two metals. The wire is under tension and slightly twists the square pin. The insulated turns of wire act as a strain relief at what would otherwise be a weak point of the joint.

Wire-wrapping was first developed at Bell Telephone Laboratories, New Jersey, United States, circa 1950.

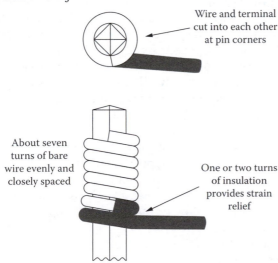

Wire and terminal cut into each other at pin corners

About seven turns of bare wire evenly and closely spaced

One or two turns of insulation provides strain relief

Figure 2.4   A wire-wrap joint.

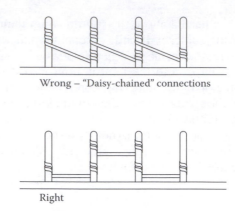

Wrong – "Daisy-chained" connections

Right

Figure 2.5   Right and wrong ways of wire-wrapping a group of terminals.

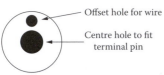

Offset hole for wire

Centre hole to fit terminal pin

End view of a wire-wrap bit

The tools required for making wire-wrap joints are more expensive than those needed for making soldered joints, but the time saved in production work soon covers the cost of the tools. The wire is stripped using a tool that ensures that the correct length of insulation is removed. The stripped end of the wire is then inserted into the offset hole of a wrapping bit, which may be part of a hand-operated or powered tool. The centre hole of the wrapping bit is then slipped over the terminal pin, and the tool is rotated evenly to wrap the wire around the pin. Most wire-wrap terminal pins have enough length for up to three joints.

One of the very significant advantages of wire-wrapping for proto-type wiring is the ease and speed with which joints can be unwrapped to allow circuit modifications. A special tool is required to unpick a joint, and any joints further out along the pin have to be unpicked first. For this reason, when a group of pins are to be connected, each wire should be at the same level on both pins connected. Figure 2.5 shows the right and wrong ways of connecting a series of pins. In the "daisy-chained" arrangement, a change to one wire can often require many other wires to be unpicked and remade. Because of the cutting action of the wire on the corners of the terminal pins, there is a limit to the number of times that a pin can be rewired. The small number of modifications likely to be needed to build a prototype are not likely to cause problems with poor joints, but continual reuse of wire-wrap IC sockets could be more trouble than the sockets are worth. The wire removed from an unpicked joint is not reusable, so in nearly all cases, both ends of a wire have to be unpicked. The unpicked wire should be carefully removed from the circuit and thrown away, taking care that fragments of stripped wire do not fall back into the circuit. An intermittent short circuit caused by a piece of loose wire can take a long time to diagnose.

One final point about wire-wrapped connections is that there is little point in trying to arrange the wires into tidy bundles: direct point-to-point wiring can be easier to inspect and check, and minimizes problems with crosstalk.

Insulation displacement

Insulation displacement is a mechanical jointing technique in which an unstripped insulated wire is pushed between the sharp edges of a forked

Tracy Kidder's classic book *The Soul of a New Machine*, first published in 1981, tells the story of a group of computer engineers working on a 1970s minicomputer and testing a prototype machine wired with wire-wrap technology.

terminal. The wire is thus held mechanically and the insulation is dis-placed by the cutting edges of the terminal, making an electrical joint. The two sides of the terminal are pushed slightly apart by the insertion of the wire and thus exert spring pressure on the conducting core of the wire. The technique is widely used for connecting discrete wires in the cabling of domestic and commercial telephone sockets, because it is fast and does not require soldering or power tools.

Insulation displacement jointing is also used in some types of cable connector, as discussed later in this chapter.

Welding

Welding is a jointing technique where two metal surfaces are placed in intimate contact and then fused together by melting both surfaces. Heat can be applied by a flame, thermal conduction, or electric heating. There are limited applications of welding in electronics, except in the bonding of ICs to their packages. External connections from bonding pads on an IC are made by attaching fine gold wires using either ultrasonic welding or thermocompression bonding, as described in the next chapter.

Crimping

A fifth jointing technique used in electronics (and much more exten-sively in electrical engineering for heavier currents) is crimping. A crimped connection is made by crushing a special terminal onto a wire of the correct size using a purpose-made tool. The wire is gripped mechanically by the crushed terminal. Electrical contact depends on the mechanical integrity of the joint. Unlike a wire-wrapped or sol-dered joint, the electrical connection is not gas tight and can therefore be prone to corrosion. Crimping is an especially useful technique for jointing unsolderable wires such as aluminium, and for rapid wiring assembly in production.

Crimping is very widely used in the automotive industry for rapid assembly of vehicle wiring harnesses.

Discrete wiring

Modern electronic product designers tend to avoid using discrete wiring in favour of printed-circuit interconnection because of the high cost of hand jointing and the likelihood of errors. Some wiring is nearly always needed, however, especially in larger systems.

A **wire** is a single or multistranded conductor with or without insu-lation, whereas a **cable** is a collection of wires or conductors bound together, possibly with some overall insulation or other protection. **Equipment wire** used in electronics is normally made of copper or tinned copper with either polyvinyl chloride (PVC) or polytetrafluo-roethylene (PTFE) insulation. PVC-covered wire can be used at up to 70°C. Above this temperature, PTFE-covered wire must be used. The size of a wire can be stated either as a cross-sectional area (in mm^2) or as the number of strands followed by the diameter of each strand. Thus "7/0.2" represents seven strands of 0.2 mm diameter. IEC 60228 speci-fies standard wire cross-sections in mm^2.

You may still find wire sizes quoted in British Standard Wire Gauge (SWG) or American Wire Gauge (AWG). You will need a table of wire gauges to find out the cross-sectional area.

For many electronic purposes, the voltage and current ratings of the lightest equipment wire are far greater than the requirements of the circuit (Table 2.1). A 7/0.2 PVC-covered wire, for example, is rated at 1.4 A (this is fairly small wire). This rating is conservative (the wire

Table 2.1   Current ratings of copper equipment wire

Construction	Cross-section area (mm²)	Current rating (A)
(a) PVC insulated		
7/0.2	0.22	1.4
16/0.2	0.5	3.0
24/0.2	0.75	4.5
32/0.2	1.0	6.0
(b) PTFE insulated		
7/0.15	0.12	3.5
7/0.2	0.22	6.0
19/0.16	0.38	9.0

could carry a greater current without overheating) to allow for the possibility of several wires being bundled together in a confined space. A 7/0.2 PTFE-covered wire is rated at 6A because the insulation can withstand higher temperatures caused by self-heating. As the following worked example shows, however, there will be a considerable voltage drop along wire of this cross-section carrying 6A.

Worked Example 2.1

Calculate the voltage drop along 500 mm of 7/0.2 PTFE-insulated copper equipment wire carrying a current of 6A. The resistivity of copper is 1.7×10^{-8} Ωm.

Solution   The cross-sectional area, A, of the wire is $7\pi \times (0.1)^2$ mm² or 0.22 mm². The resistance of a length, l, resistivity, ρ is

$$R = \frac{\rho l}{A}$$

Hence the resistance of the 500 mm length is

$1.7 \times 10^{-8} \times (0.5)/0.22 \times 10^{-6}$, or about 40 m$\Omega$

From Ohm's law, the voltage drop is about 240 mV. If this wire is used to connect a power supply to a load only half a metre away, the voltage at the load will be 0.48 V less than the voltage at the terminals of the power supply (there will be a drop of 0.24 V in each conductor) if the full-rated current of the wire is drawn.

This solution has ignored any change in resistance due to self-heating of the wire that would increase the voltage drop.

Cables

Cables are used mainly for signal and data transmission and for interconnecting subsystems within an electronic system. Table 2.2 summarizes the main types of cable used in electronic engineering, and these

Table 2.2 Common cable types

Type	Construction	Applications
Screened	One or more wires with an overall metal braid or helical screen and insulation.	Low-power signal transmission at up to audio frequencies.
Twisted pair	Two wires insulated and twisted together, covered with overall sheath, and possibly screened.	Signal transmission at up to 100 MHz.
Coaxial	One solid or stranded conductor surrounded by dielectric, metal braid, and outer insulation.	Signal transmission at up to 1 GHz.
Twin feeder	Two conductors laid parallel about 10 mm apart, insulated, and separated by a web of plastic.	Radio receiver antenna downleads.
Ribbon	10 to 50 stranded conductors laid parallel and coplanar, covered and separated by insulation.	Parallel logic interconnection in microprocessor and computer systems.

are illustrated in Figure 2.6. Multicore cables can combine these types within one cable for special applications. Each type of cable has its own range of uses and its own characteristic parameters.

There are only a few applications in electronic engineering for cables of the type known as flex. These consist of several insulated wires with overall insulation, such as three-core mains flex used to connect mains-powered equipment to a mains outlet (this is one of the few applications). The reason for this is that cables are designed to carry electromagnetic signals and the cable must either exclude unwanted signals present in the surroundings (interference) or else prevent energy escaping from the cable (and causing interference elsewhere). A screened cable is intended to prevent pickup of unwanted signals. The wire or wires within the cable carry the signal (typically from a microphone or other transducer) and are surrounded by a metal screen wound helically or woven from bare wires in the form of a braid. The screening is effective only against electric fields and high-impedance electromagnetic fields (with a strong electric component). Magnetic fields and low-impedance electromagnetic fields (with a strong magnetic component) cannot easily be screened against. The effect of a magnetic field on a cable can be reduced, however, by twisting a pair of wires together. This means that currents induced in the wires by a changing magnetic field tend to cancel because each twist of the wires reverses the polarity of the wires relative to the field. A twisted-pair cable can be used to transmit

Chapter 8 discusses electromagnetic effects in greater detail. Carter (1992) discussed electrostatic and magnetic screening.

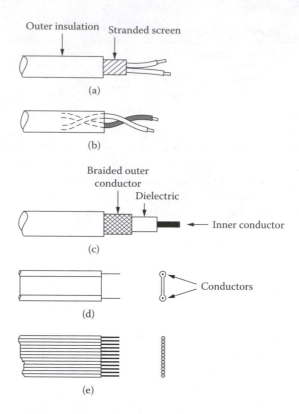

Figure 2.6   Common cable types: (a) screened, (b) twisted pair, (c) coaxial, (d) twin feeder, and (e) ribbon (insulation displacement).

frequencies of up to 100 MHz, but energy will be radiated at megahertz frequencies and a coaxial cable should properly be used for the higher frequencies. Significant applications of twisted-pair cables for high-frequency signals include 100 megabit/s (Mbit/s) Ethernet, and digital subscriber line (DSL)/asymmetrical digital subscriber line (ADSL) (which is carried over twisted-pair telephone cables). Coaxial cables will carry signals down to zero frequency, but their main use is for transmission of radio frequency (r.f.) signals at up to 1 GHz.

The theory of transmission lines is covered by Carter (1992). Any cable that is longer than the wavelength of the signals being carried must be regarded as a transmission line. Cables designed to carry signals of frequency higher than audio frequencies therefore have characteristics which include transmission-line parameters. The two most important characteristics are the characteristic impedance, Z_0, which is typically 50 to 150 Ω and the attenuation, α, which is usually stated in dB per metre, 100 m, or km (the frequency must also be given, since α depends on frequency). The characteristic impedance is independent of the length of a cable. Three other parameters commonly stated for a cable are the capacitance per metre (typically < 100 pF), the maximum working voltage, and the operating temperature range.

Ribbon cables are widely used in microprocessor and computer systems to carry parallel logic signals over distances as long as 5 to 10 m, but typically much shorter. Their main advantage over conventional multicore cables is that they can be mass terminated: a connector can

Table 2.3   Common types of cable connectors

Type	Construction	Features/applications
IEC 60320 C14	Three-pole male body with female contacts, female sockets with male contacts.	10A rating mains connector, sockets available with integral filters.
BNC 50/75 Ω	Coaxial bayonet.	Instruments, general screened and coaxial connector.
"D" type	9- to 50-way connectors with contacts in two or three rows available for soldering, wire-wrapping, ribbon cabling, and PCB mounting.	Widely used multipole connectors, 25-way version used for "RS-232" data transmission, 9-way for serial ports on computers.
RJ11/RJ45	Plastic shell with 4 or 8 sprung-finger contacts and latch.	Telephones, local area networks (LANs, or Ethernet).

be fitted to the cable in one operation taking less than a minute. If a ribbon cable is carrying high-speed digital signals, each conductor should operate as a transmission line. For this reason it is common practice to earth alternate conductors. If an even number of signals, n, is to be carried, as is normally the case, the cable should have $2n + 1$ conductors so that all signal conductors have an earth on both sides. This ensures that the characteristic impedances of all the signal conductors are equal.

Connectors

Connectors, or plugs and sockets, are used in electronic products to make electrical connections that can be easily disconnected and reconnected. They are used to connect external cables to equipment and are also fitted internally to allow subassemblies (such as PCBs) to be disconnected and removed easily for repair. One can therefore classify connector types into two main categories: cable connectors, suitable for making external connections to equipment; and wiring connectors, designed for use internally.

Table 2.3 lists a few common types of cable connectors and is limited to widely used, standardized designs. Some of these are illustrated in Figure 2.7. There are many types of proprietary connector, especially for use with multicore cable. Most cable connector types are keyed or polarized mechanically in some way so that they can be connected in one position only. Connectors designed to be attached to the end of a cable are called free connectors and incorporate some form of strain relief to grip the cable mechanically so that tension in the cable is not transmitted to the electrical joints inside the connector. Fixed connectors are designed for mounting on a panel or PCB.

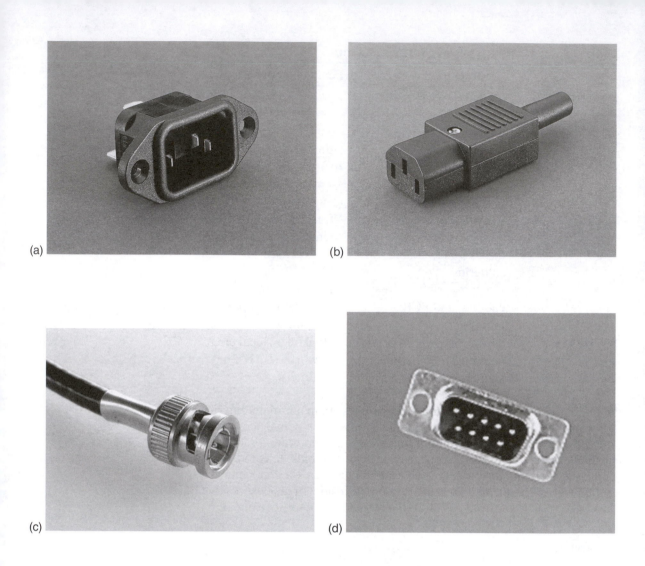

Figure 2.7   Five common types of cable connectors: (a) IEC 60320 type C14 3-pole 10 A mains connector, (b) inlet socket, (c) BNC 50Ω coaxial, (d) 9-way male "D"-type, and (e) RJ45 Ethernet jack socket (with integral light-emitting diodes [LEDs]). (Courtesy of (a,b) Bulgin Components; (c) Jonas Bergsten; (d) Amphenol-Socapex, France; and (e) Amphenol, Canada.

The contacts within a connector are referred to as male or female. When two mated connectors are separated, the live side of each circuit, if any, should be on the female contacts. The male contacts (which are accessible) should be electrically dead. In nearly all cases, this is a positive safety requirement and is one of the consequences of the widely accepted safety rule that live parts shall not be accessible.

Some of the factors to be considered when choosing a connector are: the electrical ratings and characteristics such as maximum working voltages and currents, contact resistances, insulation resistance, and transmission line parameters; the temperature ratings and intended operating environment; the reliability and life of the connector; cost; and tooling needs for making the electrical connections. The transmission line parameters are applicable only to r.f. connectors and will include the characteristic impedance and the voltage standing wave ratio (VSWR).

When an electromagnetic wave travelling along a cable meets an electrical discontinuity such as a connector, some of the wave energy is reflected and sets up a standing wave. The VSWR is a measure of the amount of reflection from the discontinuity. Chapter 7 of Carter (1992) discusses VSWR and other transmission-line parameters, such as the characteristic impedance.

The environmental conditions under which a connector will be working are most important. Connector types capable of use out of doors (splashproof or waterproof) are much more expensive than types intended for indoor use, because of the complexity of the waterproof seals and the need to protect the connector from atmospheric corrosion. The life of a connector is influenced by the number of mate–unmate operations: a heavy-duty connector is designed to withstand the wear and tear of frequent use, while other types are designed for occasional mating and unmating only. Many cable and connector types require special tooling to prepare the end of a cable and to make the electrical connections to the connector. Additionally, some training and skill are needed if a good-quality termination is to be made. A possible solution to this problem is to buy-in terminated cables from a specialist cabling contractor. Some common cable and connector configurations are available commercially as ready-made cable assemblies. This is especially true for mains voltage equipment leads fitted with IEC 60320 connectors at one end, and national mains plugs at the other end, and also for telephone and local area network cables with RJ-11 and RJ-45 jacks.

Wiring connectors are somewhat less standardized than cable connectors, and different manufacturers' products are often not interchangeable. They are used for making connections to the edges of PCBs (edge connectors), for attaching flying leads to PCBs, and for connecting ribbon cables. (Some ribbon cable connectors are suitable for making external connections, but in general their use is confined to interior interconnections.) Many wiring connectors are attached to wires by crimping rather than soldering, because crimping is a much quicker operation in production.

Ribbon cable connectors make electrical contact with the cable conductors by insulation displacement jointing, as discussed earlier in this chapter. A typical proprietary contact design is illustrated in Figure 2.8a. All the connections within the connector (up to 50) are made simultaneously in a single pressing operation, making this type of cable and connector an economical and virtually error-free method of interconnecting logic PCBs. Ribbon cable connectors are also available for connection to dual-in-line (DIL) IC sockets, as shown in Figure 2.8b.

Where individual wires have to be connected, push-on terminals (of the type commonly used in automobile wiring) or screw terminals are

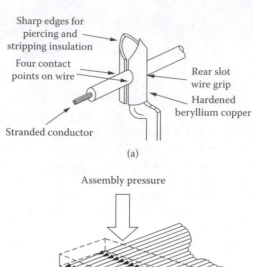

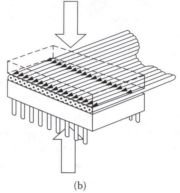

Figure 2.8   Insulation displacement connectors: (a) contact arrangement and (b) DIL plug connector assembled to cable. (Courtesy of Thomas & Betts Ltd. Design covered by U.S. Patent 3.964.816.)

often used. Some screw terminals are specifically designed to accept bared wires (tinned with solder if multistranded); others are designed to take a spade or eyelet terminal crimped to the end of a wire.

Printed circuits

General references for this section are Scarlett (1984) and Edwards (1991).

Printed circuits are used in almost all application areas of electronic engineering. Rigid printed circuit boards account for the majority of applications, providing both mechanical mounting and electrical interconnection for components. Flexible printed circuits are also popular as a substitute for discrete wiring.

Not all PCBs have electronic components mounted on them, and a significant use for PCBs is to provide interconnections among other PCBs, particularly in card cages or card racks where a number of boards slide into a frame and connect with a backplane that provides electrical interconnection among the boards. Another possibility is to use a PCB as an electrical and mechanical base on which to mount subboards containing functional blocks of circuitry of standardized design. This is known as the motherboard–daughterboard technique, and it is used both to permit modular design and to simplify manufacture by allowing easy assembly of multiple PCBs into a small space.

A prototype PCB can be manufactured by computer-controlled milling, in which copper is removed by a cutting tool. This avoids the use of chemicals.

Printed circuits are usually manufactured by chemical etching and electroplating processes. The patterns of conductors, or tracks, on a printed circuit are defined photographically from a master photographic film. The film itself is usually made by photoplotting from computer-aided design (CAD) software. For some very specialized applications, the film may be made by photographing manually prepared artwork. Before the days of cheap computers, this was the normal method for preparing PCB films.

Rigid printed circuit boards

There are three types of rigid printed circuit boards shown diagrammatically in Figure 2.9. Single-sided boards with a conductor pattern on one side only are the cheapest type. They are mainly used for very low-cost, low-component-density, consumer applications such as portable radios. Double-sided boards can carry a greater density of circuit interconnections and are much more common than single-sided boards. Double-sided boards normally have plated-through holes (PTHs), so that interconnections between one side of the board and the other are made during board manufacture. Through-hole plating is a process by which copper is deposited on the inside walls of holes drilled through the board. A thin layer of copper is first deposited by electroless plating onto the nonconducting board material. The copper thickness is then

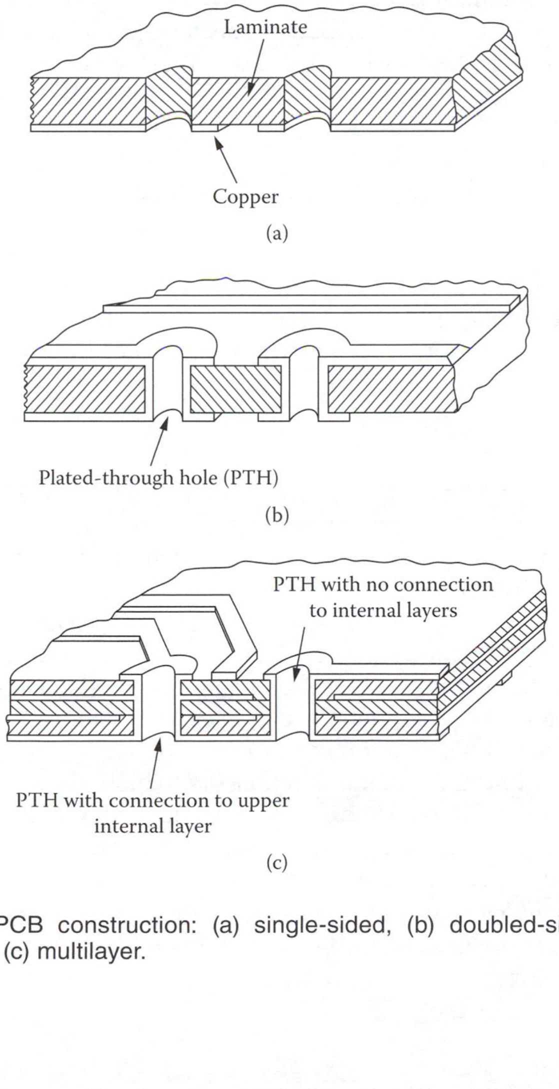

Artwork is a term used in the printing and publishing trades for image material to be reproduced photographically.

Figure 2.9   PCB construction: (a) single-sided, (b) doubled-sided plated-through, and (c) multilayer.

built up by electroplating. Double-sided boards without plated-through holes were common in the 1980s because they were cheaper to manufacture, but they are now too expensive to assemble because link wires or pins have to be soldered through some of the holes to make connections between one side of the board and the other, and through-hole plating is now a much cheaper alternative. Multilayer boards have internal layers of conductors as well as the conductors on the outer faces of the board. Connection to the internal layers is by PTHs. The manufacturing process for multilayer boards is more elaborate (and therefore more expensive) but makes possible higher-density boards than could be achieved otherwise. Multilayer boards also offer better electrical performance, for reasons that are covered in Chapter 8. They are used almost universally in computers.

Multilayer boards can be fabricated with more than 20 conductor layers, although fewer than 10 is a more usual figure.

Flexible printed circuits

Flexible circuits can be made in the same configurations as rigid boards, including through-hole plating. Multilayer flexible circuits, however, tend to have only a few layers because flexing of the board puts stress on the copper layers. Flexible circuits may be used instead of discrete wiring or to achieve a compact assembly of circuitry by folding a circuit into a small volume, which can be especially important in some military applications such as missiles. A further significant use for flexible circuits is in applications where mechanical movement must be allowed and, in these applications, the flexible circuit will be subjected to repeated flexing throughout its life. A common example is in computer hard disk drives, where the moving head is connected via a flexible circuit to the main (rigid) circuit board. Components may be mounted on a flexible circuit, but the use of larger components may require a flexi-rigid board fabricated in one piece with some sections flexible and some rigid. Flexible circuits are usually based on a polyimide plastic film with copper layers. Unlike rigid boards, the outer faces of the circuit are of polymer, not copper, to prevent delamination of the copper when the circuit is bent.

Component mounting

There are two methods of mounting components on a PCB, as illustrated in Figure 2.10. Through-hole mounting, in which component leads pass through holes in the board, is a technique that has been used from the advent of printed wiring, but has now been superceded for many applications by surface mount technology in which component leads or pads are soldered to the surface of a PCB without passing through the board. Surface mounting was tried during the 1960s, using welding to attach IC leads to copper, but it became feasible using solder jointing in the 1980s and 1990s. The main advantage of surface mounting is the smaller size of surface mount components, which allows greater component density on PCBs. Holes are still needed, of course, to make connections between one side of the board and the other, or to connect to internal layers, but these holes can be much smaller than those needed for component leads. Surface mount technology was driven by the need to make IC packages smaller (because of the large numbers of pins) but

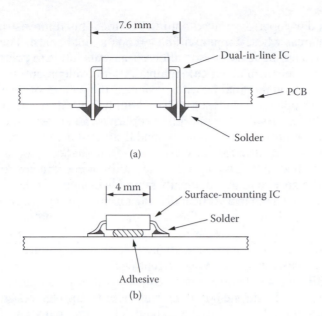

7.6 mm

Dual-in-line IC

PCB

Solder

(a)

4 mm

Surface-mounting IC

Solder

Adhesive

(b)

Figure 2.10   Component mounting methods: (a) through-hole, and (b) surface mount.

ended up contributing to the miniaturization of electronic products as smaller components were developed for surface mounting.

Rigid board materials

Most professional printed circuits are made from fibreglass-reinforced epoxy resin laminate. The fibreglass reinforcement is usually in the form of woven cloth. For very low-cost applications (such as pocket radios or cheap electronic toys), synthetic-resin-bonded paper (s.r.b.p.) may be used. Fibreglass boards are more heat resistant and more stable dimensionally than s.r.b.p., and also absorb less moisture from the air, resulting in better long-term insulation resistance. Flame-retardant grades of fibreglass board are available for applications where there is a risk of severe overheating. One advantage of s.r.b.p. is that holes may be punched rather than drilled, which reduces the manufacturing cost provided the expense of preparing the punch tool can be recovered over a large number of boards produced.

Manufacturing processes

The starting point in the manufacturing process for single- and double-sided rigid boards without PTHs is a photographic film, actual size, defining the pattern of conductors required on the board, and a piece of board material coated on one or both sides as appropriate with a layer of copper. The copper is then coated with photoresist, which is sensitive to ultraviolet light. The photographic film is then laid in contact with the resist, and the whole assembly exposed to ultraviolet light. Exposed regions of the photo-resist undergo a photochemical change, leaving an imprint of the conductor pattern on the resist. The exposed board is then chemically developed in a tank of developer or in a conveyor machine

The thickness of the copper layer is stated in micrometres (μm) or as a weight per unit area (often ounces per square foot). A common thickness is 35 μm (1 ounce-foot^{-2}).

21

in which developer is sprayed onto the board. This removes the resist, except in areas where copper will be left on the final board. These areas remain covered by resist — so called because its job is to protect these areas of copper from chemical etching. After development, the pattern of conductors on the final board is visible as the pattern of resist.

The developed board is then etched, usually in a warm solution of ferric chloride, to dissolve the unwanted copper. After etching, the remaining resist is removed with a solvent, and the board is ready for drilling.

Boards produced in quantity are drilled automatically on a numerically controlled (NC) drilling machine. Drilling coordinates for an NC drill can be generated automatically from CAD software. Drilling may also be done by hand in a vertical drilling machine, by eye (sight drilling) for a prototype board, or by using a jig for small-batch production work.

The manufacturing process for double-sided PTH boards starts with drilling of the blank board. The drilled blank has copper all over both faces to conduct electroplating current to the holes in the board. The inside walls of the holes are therefore plated with copper before the track pattern is etched onto the board. During etching, the etch resist protects the insides of the holes from the etchant. There are many variations on PCB manufacturing techniques that cannot be discussed here for lack of space.

Scarlett (1984) described many variants on the basic techniques discussed here.

Multilayer board manufacturing is a more elaborate process than double-sided PTH board manufacturing but has some steps in common. The internal copper layers of a multilayer board are etched individually, as described above for conventional boards, except that each layer of board material is thinner than conventional board laminate. Intermediate layers without copper, known as pre-preg, are not fully cured: the epoxy resin has only been partially heat-treated and is still capable of plastic flow under heat and pressure. When all the internal copper layers have been etched, the sheets of laminate and pre-preg are assembled together and bonded under heat and pressure in a press. Normally, a stack of boards is pressed at the same time, separated from each other by sheets of steel and PTFE-based plastic film.

After bonding, the board is externally similar to a double-sided PTH board: the outer faces are still covered completely with copper. The board is drilled, and the through holes are plated in the same way as in a double-sided PTH board. With multilayer boards, of course, some of the plated holes make electrical contact with internal copper layers, so the quality of the drilled holes has to be good. Finally, after through-hole plating, the outer faces of the multilayer board are etched and finished in the same way as in double-sided PTH boards.

Solder resists and legends

Most printed circuit boards are coated with an epoxy resin material called solder resist, usually dark green in colour. The purpose of this coating is to prevent solder sticking to unwanted areas of the board during mass soldering, perhaps causing short circuits, or solder bridges, between adjacent tracks. It also serves two secondary purposes: less solder adheres to the board during soldering (solder is expensive), and the coating reduces moisture absorption during the board's life, thus improving reliability. The solder resist is applied as a liquid by screen printing through a fine mesh screen.

The final manufacturing process, before assembly of components onto the board, is printing with a legend. This identifies the component positions and reference numbers on the board, the board type number, and perhaps the date of manufacture or the version number. All of this information is useful if the board has to be repaired. The legend is screen printed onto the board on top of the solder resist with an epoxy-based ink.

Printed circuit CAD

Most PCBs are designed using CAD software, and the films required for board manufacture are generated by photoplotting. In the early days of CAD software, manual design and artwork preparation were still used for high-frequency and microwave boards, but today CAD software is available even for microwave boards.

The first decisions to be made in designing a PCB are the size and shape of the board, if not already determined by the application, and the method of construction to be used. Unless there are electrical reasons for choosing multilayer construction, there is often a choice between multilayer and double-sided plated-through boards. The time taken to design a board can run into weeks, so if a double-sided design has to be abandoned partway through because of difficulty in accommodating all the connections, a considerable amount of money and time will have been wasted. On the other hand, the choice of multilayer construction adds additional material and manufacturing costs to every board produced. Of course, if a product is a development, or variant, of an existing design, the choice made for the existing design will probably be kept.

When the board dimensions and outline have been decided, the second stage in the design process is to decide how and where to position components. The component placements will be influenced by the logical structure of the circuit, by electrical requirements such as power distribution, and by the need to minimize interactions among different parts of the circuit.

Sensitive amplifier inputs, for example, would normally be positioned well away from power supply rails and output signals.

The final step in the design process is to find a path for every connection in the circuit. In practice, this stage of design interacts with component placement, as some components may have to be rearranged. Finding a path, or route, for each connection is not easy, as tracks cannot cross each other on the same side or layer. A common design for double-sided boards is to have all the tracks on one side of the board running horizontally and all those on the other side running vertically. Where a track changes from one side of the board to the other, a via hole is normally used, not a component lead hole. The process of finding paths for the tracks is called routing and is often performed automatically by CAD software, although manual routing is also possible and may produce a layout with better electrical performance, as discussed in Chapter 8. Once the routing is complete, the photographic films required for manufacture are generated in actual size by a photoplotter that exposes sheets of film with the patterns for each side or layer of the board. These films are then developed by a wet chemical process in the same way as any other photographic film.

Printed circuit assembly

Mass production of assembled and soldered PCBs is highly automated. Components are inserted into a board by high-speed automatic component insertion machines or, in the case of surface-mounted components, by machines that fix the components to the board with a drop of adhesive. Some components may have to be inserted by hand, but this is now uncommon as nearly all components are designed for automatic handling. Once all the components are in place, the whole board is mass-soldered, so that all soldered joints are made in one fast, cheap process. There are two main mass-soldering techniques in use: wave soldering and reflow soldering.

Wave soldering

Mass soldering of conventional through-hole printed circuit boards (as opposed to surface mount boards) is normally done in a wave-soldering machine. Figure 2.11 illustrates the principles of the process. PCBs loaded with components pass along a conveyor over a wave of molten solder maintained by a pump from a solder bath. The underside of the board is preheated and fluxed before the board reaches the wave. As the board passes across the wave, the underside of the board is washed with molten solder. If conditions are correctly adjusted, just sufficient solder stays on the board to make good joints, with no globules or icicles of solder on the component leads that are later cut off with rotating cutters.

Reflow soldering

Surface mount PCBs are normally mass-soldered by a reflow process. A solder–flux paste is applied to the joints, and components are stuck to the board with adhesive. The board is then heated to melt or reflow the solder by one of two methods. In vapour-phase reflow, the board is passed through a tank in which an inert fluorocarbon liquid is boiling, filling the tank with hot vapour. The vapour condenses on the board, giving up its latent heat of vapourization to the board and thus heating the solder joints. Precise temperature control is achieved because the board is heated to no more than the boiling point of the liquid. In infrared reflow, the solder paste is heated by infrared radiation from electric elements.

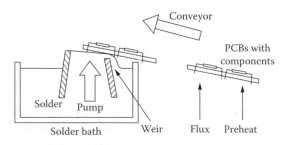

Figure 2.11   Principle of wave soldering.

Design considerations

Manufacture of a board can be made easier and cheaper by good PCB design. If a board is to be wave-soldered, for example, the direction of flow of the solder should be checked before the board is designed, and tracks on the solder side of the board laid out in the direction of flow. This reduces the chance of solder forming bridges between tracks. For the same reason, ICs should be laid out with their rows of pins across the direction of flow. If a board is to be assembled with automatic component insertion equipment, setting up the machine may be easier if all axial components are laid out on a common pitch and all integrated circuits are oriented in the same direction.

Recommendations on PCB design are given in IEC 60326-3 (1991).

Rework and repair

The ease with which wire-wrapped joints can be remade is a distinctive feature of wire-wrap technology. Soldered connections on PCBs and elsewhere may also need to be altered or remade for several reasons.

Minor faults can occur in PCB manufacturing and assembly that require some manual attention to the board. The process of correcting production faults is called rework. If a board needs attention to correct a fault that has occurred in service, the work is called repair. Similar manual techniques are applicable in either case, although the types of fault encountered may be different. Rework may involve removal of excess solder from a board, removal and replacement of an incorrect component, or addition of a component omitted during manufacture. Repair may involve reconnection of a broken wire or PCB track, replacement of failed components, and restoration of a mechanically damaged or burnt area of PCB. Some types of repair and rework on PCBs are delicate jobs for skilled craftsmen, but some limited skill at component replacement is needed by most electronics engineers doing development work on hardware.

One of the most common rework and repair operations is removal of solder to release a component from a PCB. There are two main ways of doing this, one using capillary action and one using partial vacuum. Copper braid impregnated with flux can be used to draw the solder out of a joint or hole by applying a soldering iron to the braid while in contact with the solder to be removed. Once the solder is molten, capillary action draws it into the braid and out of the joint. The end of the braid is then cut off and discarded. An alternative, less messy technique is to use a solder-sucker syringe or suction soldering iron. Here, the iron is used to melt the solder, and a sharp suck from the spring-loaded syringe or suction iron removes the molten solder from the joint. Both techniques require skill if the joint is not to be overheated, causing damage to nearby components or delamination of a PCB track. Component removal on plated-through boards can be especially difficult: when removing an IC, it is better to cut all the IC leads to release the body and then unsolder the leads one by one. It is fairly easy to pull out the barrel of a plated-through hole while removing a component lead. On a double-sided board, this may not be a problem as the new component can be soldered on both sides of the board, but if the hole

25

connects to an internal layer on a multilayer board, further repair may be impossible.

Damaged tracks on a PCB can sometimes be repaired by soldering discrete wires to the board to bridge across a damaged area, or by using self-adhesive copper strip that can be stuck to the board and soldered to the undamaged parts of the track. This is less likely to be possible on boards with very fine or closely spaced tracks. If alterations are needed to a PCB, the same methods can be used: tracks can be disconnected by cutting across in two places with a sharp knife and then carefully lifting the piece in between with the end of the blade.

Damaged areas on a PCB can be restored using epoxy resin after removal of any loose fragments or burnt areas (caused, for example, by a component being burnt out by a fault).

Further details of repair and rework techniques can be found in BS6221-21 (2001) and BS6221-25 (2000).

Case study: A temperature controller

There is no automatic method of assembling and soldering discrete wires. Wire-wrapped wiring can, however, be connected by machine in applications such as card cage backplanes.

This chapter has introduced several interconnection technologies including discrete wiring, printed circuits, and connectors. Modern electronic products are designed without discrete wiring as far as is possible because of the high cost of assembling and soldering wires. There are several different applications for discrete wiring in electronic products, summarized in Table 2.4, alongside alternative interconnection techniques that can be used for the same applications but avoid the need to solder wires by hand.

The product illustrated in Figures 2.12 and 2.13 is a temperature controller, manufactured in several versions with different types of input and output. Figure 2.12 shows an early version of the controller designed in the early 1980s, while Figure 2.13 shows a later version. Modern designs such as the controller shown in Figure 2.14 use digital displays and push-button controls for setting the temperature and are also designed from the start to have no discrete wiring and to be easily assembled.

Table 2.4   Alternatives to discrete wiring

Application	Alternative techniques
Wiring to front-panel controls, connectors, and displays.	Direct PCB mounting switches, potentiometers, and connectors. PCB fitted behind panel with ribbon cable interconnect to main PCB or right-angle mounting components fitted direct to main PCB.
Connection to transformers and power supplies.	PCB mounting components. Push-on crimped wiring connectors with PCB terminals soldered into board during mass soldering.
Inter-PCB wiring where a product consists of more than one PCB.	PCB motherboard. Board-to-board connectors. Ribbon cables connecting to PCB sockets. Flexi-rigid boards made in one piece and folded to fit into the product.

Figure 2.12   Mark 1 temperature controller.
(Product illustrated courtesy of Eurotherm Ltd.)

Figure 2.13   Mark 2 temperature controller.
(Product illustrated courtesy of Eurotherm Ltd.)

The Mark 1 version of the controller, shown in Figure 2.12, had two PCBs. The larger board included the input and output terminals at the rear edge, a transformer and power supply, temperature control circuits, and the main output circuit. The smaller board contained circuitry that varied from one version of the product to another to accommodate different types of optional second output circuit. The two PCBs were

Figure 2.14 Modern temperature controller with digital display and push-button setting. (Illustration courtesy of Eurotherm Ltd.)

connected by a wiring loom of eight colour-coded wires, making 16 joints to be soldered by hand. The wiring loom was assembled separately before soldering to the PCBs. A typical input to a controller of this type is a voltage from a thermocouple sensing the temperature to be controlled, while the output may be a thyristor circuit for controlling the power delivered to an electric furnace element. The desired temperature was set on the large wheel, and viewed through a window in the front panel. The controller is shown without its outer cover, and was typically mounted into an industrial control panel.

The Mark 2 version, shown in Figure 2.13, had no discrete wiring. It was still in production in the early 1990s, although by then it had been superceded by newer models. The smaller PCB was fabricated with edge contacts that located onto square pins staked into the main PCB. These pins were mass-soldered, with all other connections on the main board made by wave soldering. Four connections at the rear of the smaller board were made with a wiring connector pushed onto pins in the main PCB. There was thus no hand soldering needed, and the wiring loom had been eliminated. The connections formerly made by the wiring loom were then made at low cost by wave soldering with no possibility of error. This meant that there was no need for the connections to be checked: a visual inspection of the solder joint quality was sufficient. The change in wiring design together with other design changes meant that the Mark 2 version could be assembled by only five people compared to seven for the Mark 1. Those five people assembled

800 units per week. The modern design shown in Figure 2.14 has eliminated the mechanical temperature-setting wheel, and uses daughter boards to permit a range of options to be assembled simply by sliding in the appropriate daughter boards.

The objective of modern interconnection techniques in electronic product design is to minimize total manufacturing costs, and this involves a balance among material and parts costs, assembly costs, and capital costs for assembly equipment. Other factors such as parts inventories and the amount of work in progress on a production line are also important. Many of these aspects of electronic production depend on product design, and design for production is an important objective in modern electronic engineering design.

Summary

Interconnection technology is of fundamental importance in the design and manufacture of electronic products. Electrical joints in electronic systems are most often made by soldering, although for some purposes wire-wrapping or crimping is used. Soldered joints were made in the past with a low-melting-point tin–lead alloy that dissolves into the surfaces of the metals being joined. From 2006 onwards, tin–lead solder has been replaced in most applications by lead-free solders to avoid the use of lead, which is toxic. Cleanliness, flux, and sufficient heat are essential requirements for a good soldered joint. Different solders are available for different applications: the type of solder to be used should be selected with care. Wire-wrap jointing is an alternative to soldering for some applications and has advantages in ease of alteration for prototype work. The integrity of a wire-wrap joint depends on good metal-to-metal contact brought about by the pressure of the wire on the sharp corners of the terminal pin.

Discrete wiring and cabling are best avoided where possible, but are still essential in many applications. Wiring is used mainly for making internal connections, and cabling for external connections. The selection of wires and cables for a particular application requires care and attention to the characteristics of the wire or cable. Connectors are used where a connection must be easily disconnected and reconnected, and for connecting cables to equipment. The selection of a connector requires the same care as the selection of any other electronic component.

The principal interconnection technology described in this chapter has been the printed circuit, manufactured by chemical etching using photographic techniques to transfer a conductor pattern from a film to a board. Single-sided PCBs are the simplest and cheapest type. Double-sided and multilayer boards with connections between sides and to internal layers by plated-through holes are manufactured by more elaborate processes and are therefore more expensive. Printed circuit designs are usually prepared using CAD software. PCBs can be mass-soldered using a wave-soldering machine or by vapour-phase reflow.

Alterations and repairs to PCBs require skilled techniques and some special tools for removal of solder from joints.

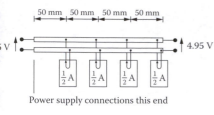

5 V

50 mm 50 mm 50 mm 50 mm

4.95 V

$\frac{1}{2}$ A $\frac{1}{2}$ A $\frac{1}{2}$ A $\frac{1}{2}$ A

Power supply connections this end

2.1   Calculate the minimum cross-sectional area of copper wire required to carry a current of 10 A with a voltage drop of less than 50 mV per metre of conductor. The resistivity of copper is 1.7 × 10^{-8} Ωm. Ignore any heating effects.

2.2   A printed-circuit backplane is to be designed for an industrial control system. Four circuit boards are to be plugged into the backplane, each drawing 0.5 A at 5 V and spaced 50 mm apart. The 5 V and 0 V power-supply connections are at one end of the backplane 50 mm from the first board. What width of track is needed in 35 μm copper if the voltage at the far end of the backplane is to be no less than 4.95 V? The resistivity of copper is 1.7 × 10^{-8} Ωm. Ignore any heating effects.

Integrated circuits 3

Objectives

☐ To emphasize the importance of integrated-circuit technology in electronic engineering.

☐ To describe the manufacturing processes used in integrated-circuit production from preparation of raw material to packaging and testing.

☐ To introduce handling precautions for semiconductor devices.

☐ To outline the range of custom integrated-circuit design methods from full-custom through gate arrays to programmable logic.

☐ To discuss briefly the importance and pitfalls of second sourcing.

The widespread application of modern electronic products is made possible, above all else, by the technology of the monolithic integrated circuit (IC), first developed during the 1960s. Integrated circuits have had a profound effect on electronic product design and utility in at least four ways. First, IC technology is inherently a mass-production technology, producing low-cost devices. Second, ICs are miniaturized circuits typically about 10 mm across or less. This makes it possible to manufacture small electronic products such as wristwatches, pocket calculators, portable phones, and MP3 players, which combine a high level of functionality with small size. Third, ICs are reliable: good-quality pocket calculators, for example, are more likely to fail through wear and tear on their keys than through the failure of their ICs. Compare this with the earliest computers built from thermionic valves, which required several failed valves to be replaced per day. Finally, the advantages just described created pressure on IC designers to reduce the power consumption of their circuits, resulting in electronic products that can be powered by small primary batteries.

The same technology that produces the IC is also used to manufacture modern discrete semiconductor devices including transistors and thyristors. Externally, these components are simple, with three or four terminals and comparatively straightforward function compared to an IC. Internally, their detailed structure can be quite elaborate, particularly for high-power devices.

From the mid-1960s onwards, the complexity of ICs as measured, for example, by the number of transistors on one chip grew exponentially. Figure 3.1 illustrates this by plotting the complexity of one manufacturer's microprocessor chips on a logarithmic scale against year of introduction. As can be seen, the number of transistors per chip doubled roughly every 18 months. Towards the mid-1980s, however, this trend began to slow down as IC manufacturers encountered the practical limits of then current IC fabrication technology. There were two contributions to the rapid doubling of IC complexity. One was a progressive reduction in the size of individual elements on the chip, and the other

Monolithic means fabricated from one piece of (literally) stone. The development of monolithic IC technology was driven by the demands of the USA's Apollo space programme to land astronauts on the Moon in the 1960s.

More detailed treatments of some of the material in this chapter are given by Morant (1990) and Sparkes (1994).

The relationship illustrated by Figure 3.1 is known as Moore's law after Gordon Moore, a founder of Intel Corporation, who first predicted exponential growth in integrated circuit complexity in a 1965 article.

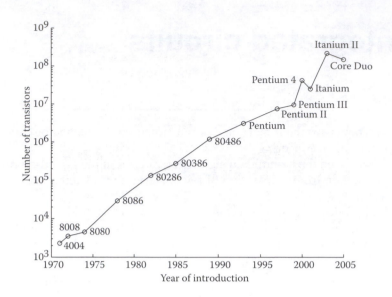

Figure 3.1   Growth of IC complexity: number of transistors on microprocessor chips plotted on a logarithmic scale against the year of introduction. (*Source:* Intel Corporation.)

was a gradual increase in the area of chips as improvements in process quality reduced the probability of defects on a chip. Innovation in circuit design also had some influence on chip complexity during the earlier years. The reduction in size of IC elements, such as transistors, caused several problems that have tended to slow down the doubling of IC complexity. One was that small circuit elements required narrower lines to be defined on the chip during fabrication but the optical photolithographic methods and materials used are limited to linewidths of no less than about 1 μm. Modern processes use deep ultraviolet light (of shorter wavelength) to define linewidths of around 50 nm. The electronic behaviour of circuit elements or devices is also affected by size reduction or scaling, so that changes in circuit design are needed as devices become smaller. The power density within a chip also increases as more and more devices are packed onto a chip and removal of heat becomes a problem.

The earliest ICs contained only a few tens of transistors or fewer than ten logic gates, and are now known as small-scale integration or SSI circuits. Later circuits, before the development of microprocessors, were known as medium-scale integration or MSI circuits and contained up to 100 logic gates or several hundred transistors. Larger chips such as 8-bit microprocessors are known as large-scale integration or LSI circuits. Chips with more than 10,000 gates, such as 16-bit and 32-bit microprocessors, were referred to as very large-scale integration or VLSI circuits. The largest chips in production in 2006 have about 200 million transistors, and the term VLSI now seems rather archaic. The possibility of fabricating a monolithic circuit covering a whole wafer, from which chips are normally separated, has been studied and given the name WSI, for wafer-scale integration, but it has never been realized commercially.

Devices may be scaled down to a few μm using simple calculations, but below this size other effects become significant and new circuit techniques are needed.

LSI and VLSI ICs tend to be digital or logic circuits rather than analogue circuits because, with only a few exceptions, analogue circuits of high complexity are not needed. Because of the complexity of modern ICs, they can only be designed with the aid of computer software, usually known as computer-aided design (CAD).

Review of semiconductor theory

Pure (intrinsic) silicon has an electrical resistivity of 2.3×10^{-3} Ωm at room temperature, which is about 11 orders of magnitude greater than that of a good conductor such as copper at 1.7×10^{-8} Ωm and at least 12 orders of magnitude less than that of a good insulator such as polystyrene at about 10^{15} to 10^{19} Ωm. Silicon is a tetravalent element in Group IV of the Periodic Table and has four electrons available for chemical bonding. Crystalline silicon is covalently bonded and has a tetrahedral lattice structure like that of diamond. The properties of silicon devices and integrated circuits depend on the ability of a silicon lattice to incorporate trivalent and pentavalent dopant atoms from Groups III and V of the Periodic Table. Group III dopants, such as boron, can contribute only three bonding electrons to the silicon lattice, leaving a vacant covalent bond or hole, which behaves as if it were a positive mobile charger carrier. Group III dopants are known as *acceptors* because the vacant bond accepts an electron from a neighbouring atom in the lattice. Silicon with an excess of acceptors is known as *p-type silicon*. Group V dopants, such as phosphorus and arsenic, have five bonding electrons, of which only four can contribute to bonding with the surrounding silicon atoms. The fifth electron is thus relatively free to move around the lattice and contribute to conduction. Group V dopants are known as *donors* because of this addition of a mobile electron to the lattice. Silicon with an excess of donors is known as *n-type silicon*. Doped silicon is known as an *extrinsic semiconductor* because its electrical properties are dominated by the effect of the dopant atoms.

A typical IC chip is about 5 to 15 mm across by 0.75 to 1 mm thick, and is cut from a wafer of semiconductor material containing hundreds of chips (Figure 3.2). Components within the integrated circuit such as transistors, diodes, and, to a limited extent, resistors and capacitors are built from regions of extrinsic semiconductor formed in the top 10 to 20 μm of the chip by incorporation of dopant atoms into the silicon crystal. Figure 3.3 shows how a p–n junction diode and an n–p–n junction transistor can be formed. The bulk of the chip shown has been doped to form a p-type substrate. An electrical connection to the substrate is made at the base of the chip for reasons that will become clear in a moment. Regions of n-type semiconductor have been formed by incorporating sufficient dopant atoms to create an excess of mobile electrons over the holes created by the p-type dopants. Within the n-type regions, further p-type and n-type regions have been formed by adding further dopants. The diode and transistor structures can be seen from the diagram to correspond with the conceptual structures shown. There are also some other p–n junctions present in the integrated circuit, formed by the n-type collector region of the transistor and the p-type substrate, and by the n-type region around the diode and the substrate. To maintain electrical isolation between the diode and the transistor, the substrate must be tied to the most negative potential in the external circuit so that the collector-to-substrate junction is reverse-biased. The surface of the chip is covered with silicon dioxide, which is an electrical insulator, and connections to the terminals of the diode and transistor are formed from a metallization layer, or layers, which is deposited over the silicon dioxide and makes contact with the underlying devices through holes, or windows, in the

The discussion in this chapter is mainly in terms of silicon because this element accounts for the majority of ICs produced today. A detailed discussion of semiconductor physics and the theory of semiconductor devices is outside the scope of this book. Till and Luxon (1982) have given a more detailed treatment of most topics in this chapter.

Two atoms bonded covalently each contribute one electron to the bond. Bonding is discussed by Anderson et al. (1990).

The majority of ICs fabricated today contain complementary metal-oxide semiconductor (CMOS) logic, because CMOS offers low power and high circuit density. Despite the low power dissipation of individual gates, a chip with millions of gates can dissipate significant power when clocked at a high frequency. CMOS is a logic technology using n-channel and p-channel metal-oxide-semiconductor field-effect transistors (MOSFETs) on the same chip. Morant (1990) has given structures for MOSFETs.

Resistors and capacitors occupy more chip area than transistors. IC designers, therefore, tend to use circuit techniques that avoid the need for passive components. Ritchie (1998) has given examples of these techniques.

In practice, diodes are often formed from a bipolar transistor with the base and collector connected to form a diode-connected transistor, which requires less chip area by a factor of β than a diode of equivalent current rating. Ritchie (1998) has given examples of IC designs using this technique.

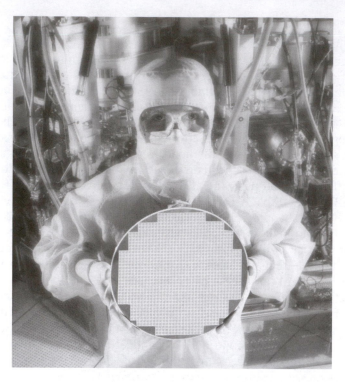

Figure 3.2   Monolithic integrated circuits: a typical modern wafer (300 mm) containing hundreds of chips. (Courtesy of Texas Instruments.)

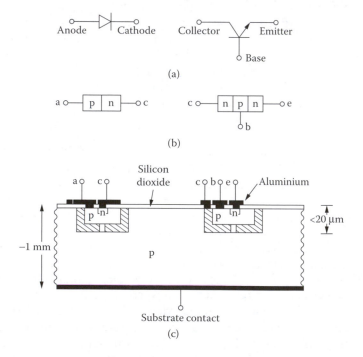

Figure 3.3   Monolithic integrated circuit structure showing possible structures for a p–n junction diode and a n–p–n junction transistor: (a) circuit symbols, (b) conceptual structures, and (c) integrated-circuit realization.

insulator. Traditionally, aluminium is used for metallization, but copper is now used for some high-performance chips. External connections to the chip are made by fine gold wires connected to bonding pads around the edge of the chip, using techniques described later.

There are other methods of isolating devices on a chip from each other and many possible geometrical arrangements of n-type and p-type regions, silicon dioxide layers, and metal layers. The geometry of IC devices and the sequence of fabrication steps required to create them are known collectively as a process. Development of a process can be a lengthy and expensive undertaking, and precise details of proprietary processes are not released by manufacturers. There are, however, only a few major steps in a fabrication process, which are combined and repeated in various sequences (up to 300 steps) to build up the desired IC structure. These process steps consist of doping processes to incorporate dopant atoms into the silicon lattice to form regions of n-type and p-type semiconductors of defined depth, lateral geometry, and dopant concentration; crystal growth to build up new layers of silicon; oxide growth to form silicon dioxide layers; lithography to transfer images to the silicon; and etching to remove regions of silicon or silicon dioxide. The entire volume of an IC chip must be a single crystal of silicon, so the dopant atoms must either be incorporated into an existing crystal lattice or be included as the lattice grows.

Morant (1990) has given examples of the sequence of steps required to fabricate ICs using a typical CMOS process.

Integrated-circuit fabrication

IC fabrication is a mass-production activity. Dozens of wafers are processed simultaneously through many of the fabrication processes, and each wafer contains hundreds of chips. Much of the processing is carried out at high temperatures of up to 1200°C, although there is a trend towards lower-temperature processing for the fabrication of ICs with small-geometry devices because of undesired dopant diffusion. The dimensions of individual device features on an IC can be as little as 50 nm in current production ICs, and smaller in research designs. Devices are approximately 0.5 μm square in production chips.

Worked Example 3.1

Estimate the size of a storage cell on a type 27128 128 k bit electronically programmable read-only memory (EPROM) chip that measures 4 mm by 5 mm (1 k = 2^{10} = 1024).

Solution The chip area is 20 mm². Dividing this by 128×1024, the area of a storage cell is 153 μm². Assuming the cells to be square, they are about 12 μm across.

Very high standards of cleanliness are required in an IC fabrication facility (Figure 3.4), because dust particles are much larger than the dimensions of device features and the image of a dust particle reproduced photolithographically on an IC may obliterate several devices, rendering the chip useless. A very high standard of air filtration and operator cleanliness is therefore essential. Operators must wear special

Clean room air quality is defined by ISO 14644 in terms of the number of particles of diameters greater than 0.1, 0.2, 0.3, 0.5, 1, and 5 μm per cubic metre of air. The highest grade of clean room (Class 1) must have no more than 10 particles greater than 0.1 μm in diameter, and no more than 2 greater than 0.2 μm, per cubic metre of air. Humans typically shed 250,000 0.5 μm particles per minute even at rest (over the whole body) and over 1,000,000 when moving about.

Figure 3.4   Part of an integrated circuit fabrication facility.
(Courtesy of Texas Instruments.)

hooded overalls to reduce contamination with skin particles. Smoking is not permitted, even during breaks outside the clean room, because of the quantity of particles exhaled for some time after smoking. Cosmetics are also banned because of the danger of particulate contamination. There is a trend towards automation in modern IC fabrication facilities to reduce the number of operators required and therefore improve cleanliness.

IC fabrication facilities are very expensive to set up and are expensive to operate, so they must produce large quantities of ICs, or ICs of high value, to be economically worthwhile.

Preparation of silicon wafers

Silicon occurs naturally as the second most abundant element after oxygen in the Earth's crust, making up about 25% by weight of the crust in the form of silicates and silica (SiO_2). Silica is found in a variety of forms including flint and quartz. Naturally occurring high-purity silica sand is the starting material for the preparation of device-grade silicon for IC manufacture. Silica is reduced to silicon by reaction with carbon in an electric furnace giving silicon and carbon monoxide:

$$SiO_2 + 2C \rightarrow Si + 2CO$$

The impure silicon is then converted to trichlorosilane ($SiHCl_3$) by reaction with hydrochloric acid:

$$Si + 3HCl \rightarrow SiHCl_3 + H_2$$

and the trichlorosilane is then purified by distillation and converted back to polycrystalline silicon by reaction with hydrogen. For device-grade silicon, typical residual impurity concentrations for Group III and V elements are less than one part in 10^9. Single-crystal silicon is required for IC fabrication. There are two methods of fabricating single-crystal ingots of silicon: the float-zone process and the Czochralski process. In the float-zone process, a polycrystalline ingot is converted to

Trichlorosilane is a colourless liquid with a boiling point of 33°C at a pressure of 758 mm of mercury. The distillation is carried out at reduced pressure because trichlorosilane is unstable and cannot be distilled at atmospheric pressure.

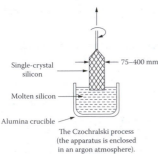

Single-crystal silicon

75–400 mm

Molten silicon

Alumina crucible

The Czochralski process
(the apparatus is enclosed
in an argon atmosphere).

single-crystal form by heating a small zone using radio frequency heating. The molten zone is passed up the ingot, and single-crystal silicon forms behind the molten zone. The float-zone process can also be used as a refinement or purification technique because impurities tend to remain in the molten zone. Most silicon for IC manufacture is made by the Czochralski process, in which a single crystal ingot is formed by slowly withdrawing a rotating seed crystal from a crucible of molten purified silicon. The crystal orientation of the silicon is determined by the orientation of the seed crystal and is carefully chosen and indicated by grinding a flat along one side of the ingot. The crystal orientation of the wafers cut from the ingot is important when the wafers are scribed and broken into chips, as cleavage occurs more easily along some crystal planes.

Till and Luxon (1982) discussed the importance of crystal orientation. For a general introduction to crystalline materials and crystal structure, see Anderson et al. (1990).

Ingots of up to 400 mm diameter can be grown. They are sawn into thin discs or wafers with a diamond saw. Small wafers of 75 to 100 mm diameter are about 0.5 to 1 mm thick. Larger sizes have to be thicker to prevent warping during processing. The surface of a sawn wafer is rough and damaged by the sawing process. The damaged layer is removed by lapping, and the wafer is then chemically etched to leave an optically smooth mirror finish. The processed wafers are inspected for flatness because later processing depends on the projection of images onto the wafer surface, and any significant deviation from flatness will cause loss of definition in the image transferred to the wafer.

Lapping is a surface-finishing process using a fine abrasive paste.

Epitaxial growth

Some semiconductor fabrication processes require the addition of extra layers of silicon on the surface of the wafer. Added silicon is known as an epitaxial layer, and it must have the same crystal structure and orientation as the wafer itself and be grown as an extension of the wafer so that no crystal boundary exists between the original surface and the new layer. An epitaxial layer might be grown after creation of a doped region in the original substrate to form a buried layer, or to form regions of different doping type or concentration to the substrate. Epitaxial layers are typically less than 20 μm thick.

Epitaxial silicon can be grown on an insulating substrate with a suitable crystal structure. An important example is sapphire (Al_2O_3) used in the silicon-on-sapphire (SOS) process for CMOS circuits.

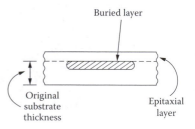

Worked Example 3.2

How many silicon atoms make up a 20 μm layer?

Solution   A crude approximation to within a factor of 2 or 3 can be calculated from the density and atomic weight of silicon, and the proton–neutron mass. (The mass of electrons in an atom is negligible.) The density of silicon is 2300 kg m^{-3}, the atomic weight is close to 28, and the masses of the proton and neutron are about 1.7×10^{-27} kg. A silicon atom has a mass, therefore, of (28) 1.7×10^{-27} or 4.8×10^{-26} kg. The number of atoms in a cubic metre of silicon is 2300/4.8×10^{-26} or 4.8×10^{28}. Assuming the atoms to be packed cubically (which they are not), there would be $\sqrt[3]{4.8 \times 10^{28}}$ atoms along each edge of the cube. The number of atoms across the thickness of a 20 μm epitaxial layer is thus $\sqrt[3]{4.8 \times 10^{28}} \times 20$ μm or about 70,000, a surprisingly small number compared to the enormous numbers of atoms in bulk material.

Silicon atoms are added to the substrate surface in a reactor at a temperature of 800 to 1100°C. The required silicon atoms can be produced by the pyrolytic decomposition of silane (SiH_4) at around 1000°C.

$$SiH_4 \rightarrow Si + 2H_2$$

This is the overall reaction. The reaction

$$2SiCl_2 \rightarrow Si + SiCl_4$$

takes place on the silicon substrate after production of $SiCl_2$ in the gas stream by the reaction

$$SiCl_4 + H_2 \leftrightarrow SiCl_2 + 2HCl$$

or by reduction of silicon tetrachloride ($SiCl_4$) by hydrogen

$$2SiCl_4 + 2H_2 \rightarrow Si + SiCl_4 + 4HCl.$$

Dopants can be included as an integral part of an epitaxial layer by adding traces of gases such as diborane (B_2H_6), phosphine (PH_3), and arsine (AsH_3) to the gas flow through the reactor. Careful control of the gas concentrations is essential if the number of crystal defects in the epitaxial layer is to be minimized.

Oxide growth

One of the most frequent steps in many IC fabrication processes is the formation of a layer of silicon dioxide (SiO_2) on the wafer surface. Silicon dioxide is an excellent dielectric, and it can be used as an insulating layer within a device such as an insulated-gate field-effect transistor (IGFET, or MOSFET) or to isolate an epitaxially grown region of semiconductor from the substrate. A final layer of silicon dioxide (apart from the metallization layer, which is described later) can passivate the surface of an IC to protect it from atmospheric contaminants. The most significant application of silicon dioxide in IC fabrication is as a masking layer that is etched to define regions to be doped.

The thicknesses required for these two purposes are very different: typically 20 nm (0.02 μm) for an IGFET gate and up to 1 μm for an isolation layer.

A silicon dioxide layer can be produced by a chemical reaction between the wafer surface and either oxygen or steam, or by a deposition process similar to epitaxial growth. Thermal oxidation requires a temperature of 800–1250°C controlled to within ±0.5°C or better together with careful control of the oxygen concentration (by using, for example, oxygen–nitrogen mixtures) and the time of processing to within 5 to 20 seconds. Because oxidation is a reaction between oxygen and the silicon surface, the reaction rate reduces as the thickness of oxide increases. Growth of a 1 μm isolation layer can take over an hour, whereas the formation of an IGFET gate layer may take only a few minutes. The oxide layer forms at the expense of the underlying silicon because silicon atoms from the wafer react to form the oxide. The thickness of silicon converted to oxide is about 30% of the final thickness of the oxide layer.

Silicon nitride (SiN_3) is another possible dielectric used in IC fabrication.

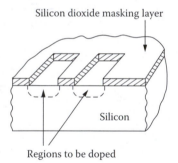

Silicon dioxide masking layer

Silicon

Regions to be doped

Thermal oxidation is carried out in quartz furnace tubes slightly larger than the wafer diameter and up to 2 m long. The wafers are held vertically in quartz boats or carriers and are processed in large batches of perhaps 100 wafers at a time.

Lithography

Lithography means, literally, "writing on stone." The techniques discussed here are in principle the same as those used for PCB manufacture and described in Chapter 2. The scale of lithographic images in IC fabrication is, however, about 1000 times smaller than those used in PCB manufacture. (Typical linewidths are 0.5 μm and 0.5 mm respectively.)

The various regions of extrinsic semiconductor, oxide, and metallization making up the devices and interconnections on a chip have to be defined in the form of an image on the wafer surface. Techniques to do this are known as lithography.

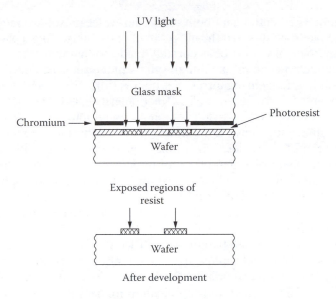

UV light

Glass mask

Chromium →   ← Photoresist

Wafer

Exposed regions of
resist

Wafer

After development

Figure 3.5   Principle of photolithography.

The earliest technique used in IC fabrication, and still very important today, is photolithography. Figure 3.5 illustrates the principle. The wafer is coated with a layer of photoresist a few micrometres thick. The resist is applied to the wafer as a drop of liquid while the wafer is spinning at high speed, ensuring that the resist is evenly distributed. The wafer is then gently baked to drive off the resist solvent. Resists are polymeric materials sensitive to ultraviolet (UV) light. Exposure to UV radiation can either cause a polymerization reaction or depolymerize the resist depending on whether a negative or positive image is required. The earliest resists for IC fabrication were of the negative image type, but for modern IC fabrication, positive resists are used because of their superior definition. After exposure, the resist is developed in a chemical solution that dissolves the unpolymerized regions of resist, leaving selected regions of the wafer coated with a tough polymer to resist chemical etchants or ion beams.

The image pattern to be transferred to the resist is defined by a photomask (or just "mask"), or reticle. These are thin quartz plates from 1.5 to 3 mm thick and originally as large as the wafer. For modern wafers of 200 to 400 mm diameter, the reticle is much smaller, and it is moved in steps (by a stepper machine) across the wafer to expose the whole wafer in a series of exposures. The image is defined on the mask or reticle by a chromium layer, which is generated lithographically from the IC design, either by photographic reduction from artwork or by the electron-beam lithography technique (described below). The earliest ICs were fabricated by contact printing, in which the chromium side of the mask touched the resist-coated wafer. The obvious problem with contact printing was damage to the mask, and to overcome this, projection printing systems were developed so that the mask could be kept away from the wafer surface. Projection printing is essential when using steppers, and a reduction lens is used so that the pattern on the chip is several times smaller than that on the mask. Accurate alignment or registration is essential, and both mask and wafer must have alignment marks to facilitate this. All the masks in a set for fabricating a particular IC design must of course be accurately made so that registration of the image defined by one mask with all others in the set is achieved.

Photolithography using ultraviolet light is limited to linewidths of no less than about 1 µm because of diffraction effects at line edges. Short-wavelength (deep) UV is used for photolithography of most modern ICs with linewidths down to 50 nm or so. An important technique that also overcomes the resolution problem is electron-beam lithography. An electron beam of around 0.2 µm diameter or less is directed onto a resist-coated surface (either a mask or a wafer) on a high-precision $x–y$ coordinate table. The resist must be an electron-beam resist, not an optical resist, and the $x–y$ table must be accurate to within fractions of a micrometre. The process is slow: full exposure of even a small 75 mm wafer can take more than an hour. Electron-beam lithography has the very significant advantage for mask fabrication of direct transfer of design information from a CAD system to the mask with no optical reduction process. It also finds application for fabrication of prototype chip designs without the expense of making a mask set, using direct-write-on-wafer imaging.

Etching

Etching is the removal of unwanted regions of material from a wafer. A typical example is cutting of holes or windows in a silicon dioxide layer prior to dopant diffusion or implantation into the regions under the windows. The regions to be etched are defined by the pattern in a resist created by a lithographic process as described in the previous section. There are two important etching methods used in IC fabrication: wet etching and plasma etching.

Wet etching was the earliest technique and is still used in production. Wafers to be etched are immersed in an acid bath and agitated to ensure even etching. Silicon is etched with a nitric acid–hydrofluoric acid mixture, and silicon dioxide with a hydrofluoric acid–ammonium fluoride solution. The amount of material removed is dependent on temperature and immersion time.

Plasma etching is a dry technique in which reactive gaseous atoms react with the exposed regions of the wafer to form gaseous reaction products that are removed by a vacuum pump. The reactive atoms are generated by breakdown of molecules in a gas heated by radio frequency electromagnetic energy.

Both wet and dry etching have the disadvantage that the etchant undercuts the resist as sketched in the margin. This effect has to be allowed for in the design of an IC layout, and it limits the packing density that can be achieved.

Diffusion and ion implantation

In order to produce regions of n-type and p-type semiconductor within a wafer, dopant atoms must be introduced into the crystal structure. Dopants can be incorporated into epitaxial layers during deposition as described earlier. If dopants are to be incorporated into an existing crystal, they

A *plasma* is a fully ionized gas consisting of electrons and atomic nuclei. The term is used here to refer to a partly ionized gas consisting of electrons, ions, and neutral atoms.

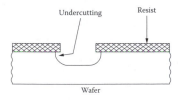

Undercutting   Resist

Wafer

can be introduced by solid diffusion or ion implantation. Both techniques require an oxide masking layer to define the regions to be doped.

Diffusion was the earliest process used for doping a wafer and takes place in two stages: predeposition and the diffusion process itself. The dopant material can be deposited on the wafer by spin coating with a liquid or by deposition from a gas in a diffusion furnace. After deposition the dopant atoms are still concentrated near the wafer surface. The second stage of the diffusion process distributes the dopant atoms to the required depth in the wafer by heating the wafer to around 1000°C in a diffusion furnace, of identical construction to the furnace described earlier for oxide growth. Separate furnaces are essential for diffusion and oxide growth, otherwise the oxide furnace will become contaminated with dopants. During diffusion, the windows in the diffusion-masking oxide layer must be sealed to prevent the dopants from diffusing out of the wafer. This can be done by regrowing oxide over the windows by using an oxidizing atmosphere in the diffusion furnace. Some predeposition processes produce a surface glassy layer that serves the same purpose. Careful control of diffusion time and temperature ensures that the dopant atoms diffuse to the desired depth, which is typically between 0.3 and several micrometres. Diffusion also occurs laterally, of course, and this must be allowed for in the process design rules.

Ion implantation is a more recent method of doping in which the dopant is fired at the wafer as an ion beam inside a vacuum chamber. The ion energy is typically between 25 and 200 keV, and can be precisely controlled so that the ions penetrate the wafer surface to a controlled depth. The total quantity of dopant introduced into the wafer can also be controlled to within ±10% or better. Typical ion doses are between 10^{15} and 10^{20} ions m^{-2}. The ion beam is broad compared to the line dimensions on the wafer, but is narrow compared to the wafer dimensions, so that the beam has to be scanned and the wafer rotated to ensure even exposure to the beam.

Implanted ions are not bonded into the silicon lattice: they occupy interstitial sites among the lattice atoms. The silicon lattice itself is also disrupted by the implanted ions as they lose energy by collision with the lattice atoms. After ion implantation, therefore, wafers must be heat-treated to repair the crystalline structure and activate the implanted dopant by incorporating the dopant atoms into the silicon lattice. Dopant atoms that are bonded into the lattice in place of a silicon atom are said to occupy substitutional sites. The heat-treatment process is known as annealing and is carried out at similar temperatures and for similar times as thermal oxidation and diffusion. During annealing, of course, diffusion occurs and the implanted dopants migrate through the lattice to some extent.

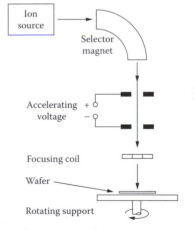

Metallization

The final stage of wafer processing is deposition of a metal interconnect to connect the individual devices on the chips and to form bonding pads for connection to the external circuit. Four or more layers of metal can be used, separated by dielectric with different conductor patterns on each layer. Windows are etched in the dielectric layers to make contact

Some interconnections may be made by diffused regions or by deposited polycrystalline silicon layers.

with the separate devices. A layer of metal, usually aluminium, sometimes with a small amount of silicon, is then deposited on the wafer, usually by vacuum deposition from vapour in a vacuum chamber. After deposition, the wafer is sintered at around 400°C. This improves the electrical connection between the deposited metal and the silicon. Modern high-speed ICs often have copper interconnects rather than aluminium. The better conductivity of copper allows faster charging of on-chip capacitances and therefore shorter rise and fall times on logic signals, which permits faster clocking.

In some devices, such as Schottky diodes and transistors, the metallization forms part of the device as well as being an interconnect. This is also true in metal-gate MOS (metal-oxide semiconductor) devices, where the gate electrode is fabricated as part of the metallization layer.

A metal–semiconductor junction is known as a Schottky junction. Schottky transistors are used in Schottky and low-power Schottky transistor–transistor logic (LS TTL) circuits.

Testing, dicing, and bonding

After metallization, all the individual ICs on the processed wafer have to be tested. Testing machines have needle probes that are pressed onto the bonding pads around the edge of the chip. There may be additional pads for test purposes that will not be connected externally. Not all the ICs on a wafer will work: those that fail the test are noted by the testing machine for later rejection. (This was once done with a drop of ink to mark the rejected chips, but modern machines simply note the coordinates of the rejected chips and remove them mechanically after dicing.) The percentage of chips that pass the test is known as the production yield. In the early production batches of a new chip or process, the yield may be quite low — many of the chips on a wafer are rejects.

Testing large digital circuits is a nontrivial operation. Wilkins (1990) has given an introduction to the problem and discusses some approaches to testable design.

The wafer is now scribed with a diamond scriber to separate the wafer into chips, each of which will be mounted into an IC package. An electrical connection to the back of the chip is necessary for most types of IC, and a common technique is eutectic bonding to a gold-plated header. Gold and silicon form a eutectic alloy at 370°C. The bonding pads on the chip now have to be connected to the external leads or pads of the package. This was once done by hand by an operator using micromanipulators working through an optical microscope, but is now carried out by automatic machinery with automated optical inspection. Gold wire is connected by ultrasonic welding or thermocompression bonding between the bonding pads on the IC and the package connections. The number of connections varies from eight for a 741-type operational amplifier to 400 or more for a modern microprocessor. Once all the connections are in place, the package can be sealed, moulded, or otherwise completed. The packaged ICs are then tested electrically again and may also be subjected to heat, vibration, pressure, or vacuum to detect weak circuits that would otherwise fail early in life.

Chips may also be used unpackaged in hybrid microcircuits or in multichip modules (MCMs). Hybrid microcircuits consist of a combination of bare monolithic ICs with printed resistors and capacitors mounted on a ceramic substrate. Till and Luxon (1982) discussed the technology of hybrid microcircuits. Multichip modules can consist of bare monolithic ICs attached to a silicon substrate with interconnects fabricated on the substrate.

Semiconductor packaging

Figure 3.6 illustrates a selection of packaging styles used for ICs, while Table 3.1 summarizes the main characteristics of the most important package types. Hermetically sealed (airtight) metal cans were the earliest form of IC package and are still available for a few linear circuits. Power voltage regulator ICs are often packaged in metal cans with a thick base similar to a power transistor package. The dual-in-line (DIL) package has been used since the 1960s for packaging both logic and linear circuits, but it was largely superceded in the 1990s by surface mount packages, although many ICs are still available in DIL packages. The DIL package is designed for soldering into through-holes on a printed circuit board (PCB) or for insertion in a socket. Plastic DIL packages are moulded onto the metal lead frame after the chip has been bonded to the leads. Ceramic DIL packages may be more reliable than plastic packages, but they are also more expensive. They can be of either the frit-seal type, consisting of two ceramic slabs cemented together with a glassy ceramic onto a lead frame, or the side-brazed type with leads brazed onto connection pads along the sides of the package. The IC chip is mounted in a cavity in the ceramic and covered with a hermetically sealed metal lid in the case of a side-brazed package, or with the top slab of ceramic in the case of the frit-seal package. The most common

See Figure 7.1b for an illustration of a power transistor package.

Plastic packages might, for example, allow ingress of moisture along the leads, leading to corrosion of the leads or the IC bonding pads.

Table 3.1 Characteristics of the principal types of integrated circuit package

(a) Through-hole mounting types, 2.54 mm lead spacing

2.54 mm is equal to 0.1 inch. Early packages were designed to Imperial (inch) dimensions except in the former USSR and Eastern Europe, where metric packages were used. Some newer package types are designed to metric dimensions.

Type	Maximum number of leads	Features
Hermetic metal can	12	Leads on pitch circle, almost obsolete
Plastic dual-in-line (DIL)	48	Low cost
Ceramic dual-in-line	64	High reliability
Ceramic pin-grid array (PGA)	> 400	Low board area for number of pins
Ball-grid array (BGA)	> 400	Designed for reflow soldering to boards

(b) Surface-mounting types, 1.27 mm lead or pad spacing

Type	Maximum number of leads or pads	Features
Plastic small outline (SO)	28	Low cost
Plastic-leaded chip carrier (PLCC)	124	Compact, low cost
Leadless ceramic chip carrier (LCCC)	124	High reliability

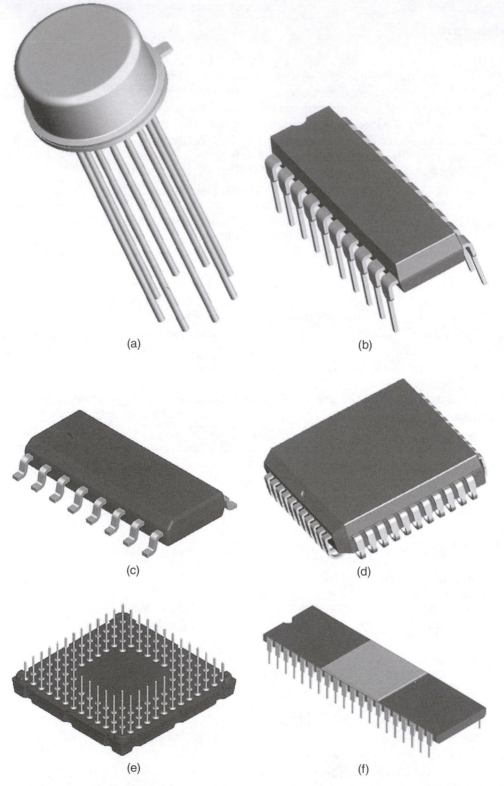

Figure 3.6 Integrated circuit packages: (a) hermetic metal can, (b) plastic DIL, (c) plastic small outline (surface mount), (d) PLCC, (e) PGA, and (f) ceramic DIL. (Courtesy of National Semiconductor. These package depictions are for example only and should not be used to design with. For current packages and accurate dimensions, please refer to the National Semiconductor Web site at: http://www.national.com/.)

application for the frit-seal package is UV-erasable memories, where the chip is visible through a quartz window in the top slab.

When DIL packages were first manufactured, ICs typically had 14 or 16 external connections, making a package about 20 mm × 7 mm. As LSI chips became available with 40 or more connections, the DIL package became unwieldy: a 40-pin DIL package measures about 50 mm × 16 mm, yet may house a chip about 5 mm square. To reduce the size of IC packages for large chips, the pin-grid array (PGA) was developed, with pins still on a 2.54 mm (0.1 inch) pitch but arranged in several rows all around or all over the underside of the package on a rectangular grid. PGAs are constructed in the same way as a side-brazed ceramic DIL package. The PGA can have up to 400 leads and yet be only 25 mm square. Larger PGAs with over 400 leads are used with sides of 47 mm or more.

At about the same time that pin-grid arrays were introduced, the technology of surface mounting was also being developed. Because several manufacturers were working on surface mounting simultaneously, several types of package emerged. The small-outline (SO) package is essentially a scaled-down DIL package with leads spaced at half the pitch of a DIL package and folded out flat rather than projecting beneath the package. The chip-carrier package has leads spaced at the same 1.27 mm (0.05 inch) pitch as the SO package, but on all four edges. Plastic-leaded chip carriers (PLCCs) are moulded in a similar way to SO packages but have J-shaped leads rolled under the package body. Leadless ceramic chip carriers (LCCs) have pads rather than leads and are similar to a side-brazed ceramic DIL package in construction. There are also quad flat packs that are similar to SO packages but with leads around all four edges of a roughly square package. Lead spacings vary, but can be as little as 0.5 mm. Flat packs are usually moulded from plastic. PLCCs and LCCs can be housed in sockets. SO packages and flat packs are not socketable. Ceramic chip carriers are not suitable for soldering to epoxy PCBs because of the difference in thermal expansion coefficient of the ceramic relative to epoxy, which can cause stress cracking of solder joints. Ball-grid arrays (BGAs) are packages designed for direct surface mount soldering by reflow — they have small solder bumps arranged in a rectangular grid on the underside of an epoxy plate. The chip is mounted on the top surface of the plate under a plastic cover. BGAs can also be mounted in sockets. Some specialized packages are used for processor chips such as the Intel Pentium, where the heat dissipation at full power is significant (50 W or more).

This point is discussed further in Chapter 10.

Handling of semiconductor devices

Many types of semiconductor devices and integrated circuits are damaged fairly easily by physical or thermal shock, overheating during soldering, and, especially, electrostatic discharge. The damage caused by mishandling is often not catastrophic — the device does not fail immediately, but is weakened by the damage, and eventually fails weeks or months later after being built into a product and shipped to an end user. Careful quality control is therefore essential on an electronics production line. Production operators must be made aware of handling

Test methods are discussed in Chapter 10.

precautions and provided with the correct tools and equipment for handling semiconductors. Completed electronic products can also be tested to reveal defects such as weakened semiconductor devices before they leave the factory.

Thermal damage

Germanium diodes, for example, find application in some analogue circuits because of their low forward voltage drop compared to silicon diodes. The 1N34A and OA90 are examples of germanium diodes that are still available.

Germanium semiconductors were the earliest solid-state devices, and they were easily damaged by overheating during soldering. Although silicon devices are used for nearly all applications today, a few germanium devices are still available and are used in specialized applications. Heat-absorbing pliers are recommended for holding the leads of these devices while soldering. Silicon semiconductor devices are more robust thermally, but they still require care if soldered by hand. Mass soldering is more easily controlled both in temperature and in time, to keep within the limits specified in a data sheet (usually under the heading "Absolute maximum ratings").

CMOS and MOS logic circuits have gate-protection diodes connected to external gate electrodes to provide a path for static charge to leak away. This may not, however, prevent damage caused by a discharge into the gate terminal from an external source.

Certain types of integrated circuits and discrete semiconductors are sensitive to damage by discharge of static electric charge. Not all device types are sensitive to damage (although none is completely immune). Field-effect devices with insulated gate electrodes are the most sensitive types. These include MOS and CMOS logic circuits and MOSFET transistors. Damage to the devices results from electrostatic discharge (ESD) into the gate electrodes, causing dielectric breakdown and perforation of the gate insulation. Electrostatic charge can build up on clothing, shoes, and floor coverings as a result of surfaces rubbing together. Figure 3.7 shows a typical ESD protected area (EPA) for electrostatic sensitive devices as recommended in an international standard. The workbench surface, compartment trays, floor mat, and operator's stool are all electrically conductive. The operator is wearing electrostatically conductive overalls and special shoes, and is connected by a wrist strap to earth. Because of the danger of electric shock in an environment of earthed conductors, the electricity supply to the workbench is isolated from the mains supply by an isolation transformer and is fitted with a residual current device (RCD). The earthing straps have a resistance of 0.5 to 1 MΩ, sufficient to conduct static charge safely to earth, but high enough to limit current flow in the presence of an electrical fault to a safe level.

Symbol denoting electrostatically sensitive devices. (Courtesy of the Electrostatic Discharge Association.)

Electrical safety is discussed in Chapter 11.

Electrostatically sensitive devices are protected in storage and in transit using special packaging materials. Dual-in-line ICs, for example, are inserted into conductive foam or kept in an electrostatically conductive plastic tube. Assembled PCBs containing sensitive devices can be placed in electrostatically conductive plastic bags for shipment and storage.

Custom integrated circuits

Morant (1990) has discussed the IC design process in greater detail.

There is a wide variety of standard integrated circuits available, both for logic and for analogue applications. If an electronic product is to be manufactured in quantity, however, it may be economically worthwhile to design a tailor-made custom IC specifically for the product. These circuits are known as ASICs, for application-specific integrated circuits. A custom-designed IC is cheaper per IC, more reliable, and of smaller size and power consumption than the equivalent circuit implemented with

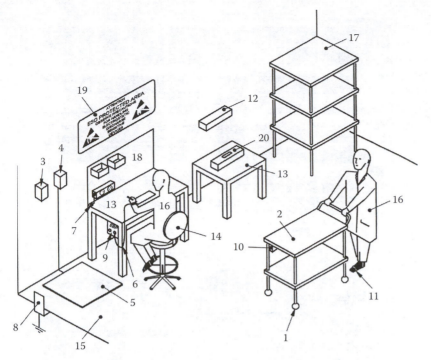

1 Groundable wheels
2 Groundable surface
3 Wrist strap tester, shall be displayed outside the EPA
4 Footwear tester, shall be displayed outside the EPA
5 Footwear tester foot plate
6 Wrist cord and wrist band (wrist strap)
7 EPA ground cord
8 EPA ground
9 Earth bonding point (EBP)
10 Groundable point of trolley
11 ESD protective footwear
12 Ionizer
13 Working surfaces
14 Seating with groundable feet and pads
15 Floor
16 Garments
17 Shelving with grounded surfaces
18 Groundable racking
19 EPA sign
20 Machine

Figure 3.7   A recommended ESD protected area for electrostatically sensitive devices. (Reproduced from IEC Publication IEC 61340-5-1. Copyright © 1998 IEC, Geneva, Switzerland. www.iec.ch.)

standard SSI and MSI components. Also, the reduced size or greater functionality (or both) of the resulting product may give the company using a custom IC a market advantage over its competitors. There are several different approaches to custom IC design, each applicable over a certain range of production volumes, because production costs per IC are inversely related to setup and design costs. The most expensive option to set up (full custom design) produces the lowest-cost ICs, but only if hundreds of thousands of ICs are to be made. Conversely, the cheaper techniques such as gate arrays are much less expensive to set up, but produce more expensive chips. For small production runs of perhaps 10,000 ICs, however, they may offer the cheapest total cost.

Full-custom integrated circuits

The most expensive form of ASIC is the full-custom integrated circuit. These are designed and manufactured in exactly the same way as standard ICs. They are justifiable only for very high-volume applications

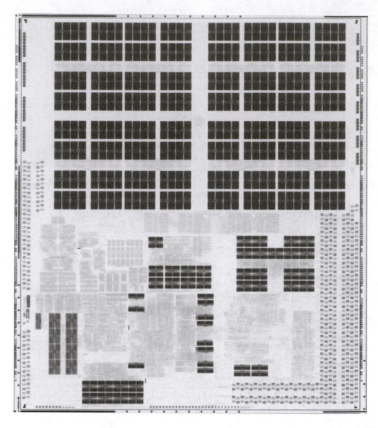

Figure 3.8 A standard-cell chip. Note the unused areas and the different sizes of cells. (Courtesy of Fujitsu Limited. Copyright © 2006 Fujitsu Limited. All rights reserved.)

Trade-offs among unit cost, setup or design costs, and manufacturing quantity occur in all fields of engineering.

(hundreds of thousands of ICs per year) because of the high cost of design. They also are the cheapest type of custom IC per unit manufactured.

Companies using full-custom ICs in their products either have their own design and manufacturing facilities or contract out the work to a semiconductor manufacturer.

Standard-cell integrated circuits

Computer-aided design (CAD) has made possible a cheaper approach to custom IC design, known as the standard-cell system. A computer database holds a library of common circuit elements such as logic gates, flip-flops, operational amplifiers, and so on stored in the form of their physical layout within the chip. The library is similar to the standard small- and medium-scale ICs used in noncustom design, but it may also include larger components such as microprocessors (known as "cores"). A standard-cell library, however, can hold many more designs than are available as standard ICs. A standard-cell chip is designed by assembling the required cells from the computerized library (using CAD software). This involves not only the logical circuit design but also the physical placement of cells on the chip. Since the shape and size of the

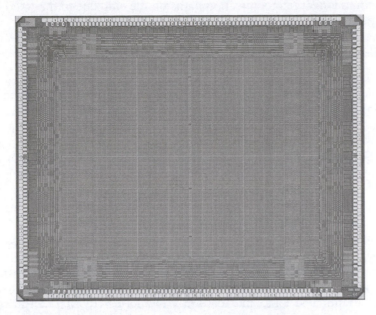

Figure 3.9   A Xilinx field-programmable gate-array chip showing a regular array of configurable logic blocks. (Courtesy of Xilinx Inc.)

cells are fixed, there will be unused areas among cells (Figure 3.8). A standard-cell design will therefore occupy more chip area than a full-custom equivalent. Production costs will be higher than for a full-custom design because there will be fewer chips per wafer, but on the other hand design costs are lower because a lot of the design work has already been done in designing the standard cells (and the cost of that work is shared among all the purchasers of the cell library).

Once designed, a full set of masks has to be fabricated exactly as for a full-custom IC, and the manufacturing process is identical.

Gate arrays

A third method of custom IC design and manufacture exists, with a lower design cost than the standard-cell system. It also differs from full-custom and standard-cell techniques in the manufacturing stages. The gate-array manufacturer designs a standard chip with a fixed layout of transistors or logic gates and flip-flops, and produces a full mask set with the exception of the metallization masks. Wafers are processed in quantity by the normal processes, but no metallization is applied. The customer designs the metal interconnect using CAD software. A custom mask is then made for the interconnect, and preprocessed wafers are metallized to the customer's design (Figure 3.9).

The setting-up costs of design and mask manufacture are comparatively low for gate arrays, so that a gate-array chip design may be viable for production quantities of only a few thousand per year. Gate arrays have several disadvantages, however, compared to standard-cell design,

Small batches of gate-array ICs may be metallized by electron beam lithography, which was described earlier in this chapter.

including longer interconnection paths on the chip due to the fixed layout of the array elements.

Programmable logic and gate arrays

The cheapest form of custom IC in terms of design costs is programmable logic. Several types of device exist, but all have in common the fact that they are manufactured in large quantities, fully packaged but uncustomized. They contain an array of basic elements interconnected by user-programmable links. The links can be fuses or insulated-gate MOS transistors that can be selectively burnt through or charged respectively to define the connectivity and therefore the function of the logic. Field-programmable gate arrays (FPGAs) are similar to the gate arrays described in the previous section except that interconnections among logic elements are programmable. Programmable logic arrays (PLAs) consist of a fixed AND–OR structure with programmable connections through a diode matrix. The same technology can be used for programmable read-only memories (PROMs) used in computers and microprocessor systems to store machine-code programs and processor microcode.

Programmable logic and read-only memories (ROMs) can also be fabricated as mask-programmed devices in which the logic function or memory contents are defined by the metallization mask. Single-chip microprocessors with on-board ROM can also be mask-programmed. Mask programming allows lower-cost, high-volume production once the logic or program design has been proven.

Any electronics manufacturer, whether using standard ICs or custom-designed parts, needs more than one source of supply to guard against component shortages caused by technical or other problems at the IC manufacturer's facilities. The semiconductor industry recognizes this need and tries to set up second sourcing wherever possible. Typically, two semiconductor companies exchange mask sets so that each can manufacture and sell the other's designs. In some cases, very popular parts such as the 741 operational amplifier or the 555 timer IC become available from five or more manufacturers. Problems can occur, however, especially with microprocessors and other LSI chips, if the different manufacturers have designed their chips independently rather than exchanging mask sets, because of detail differences among the nominally identical products.

Summary

Integrated-circuit technology is a very important part of modern electronic engineering. It makes possible low-cost, highly reliable products with a high level of functionality. Most integrated circuits are fabricated from silicon. Integrated transistors and other devices are formed within a silicon wafer by incorporation of dopants into the silicon crystal structure. The regions to be doped are defined by lithography and etching of a silicon dioxide layer grown on the surface of the wafer. Dopants can be introduced by diffusion or ion implantation. Additional silicon may be built up on a wafer by epitaxial growth, in which silicon atoms are deposited on the wafer from a gas. The epitaxial layer is an extension of the existing crystal structure of the wafer. Interconnections among devices on an IC may be made by a deposited metallization layer. Throughout all of these processes, whole wafers containing hundreds of ICs are processed together often in batches of dozens of wafers. There are several different styles of IC package, each with its own advantages and disadvantages. These can be broadly divided into through-hole or surface-mounting types and into plastic and ceramic types. Surface-mounting types are smaller than equivalent through-hole types, and ceramic packages are more reliable and expensive than plastic packages.

Semiconductor devices and ICs are sensitive to thermal and electrostatic damage and correct handling precautions must be taken in a production environment to reduce product reject rates.

ICs for custom applications may be designed in the same way as standard production ICs provided large quantities are required, or by assembling a design from predesigned standard cells, or by designing a metallization layer to interconnect a mass-produced, but unmetallized, gate array. For small batches of custom circuits, field-programmable gate arrays or logic arrays may be used.

Problems

3.1 At the time of publication of the first edition of this book in 1987, VLSI ICs with 10^6 transistors per chip were reaching production. Using the smallest discrete transistors available, which are surface mount devices measuring about 2 mm × 3 mm, what area of PCB would be required to realize the equivalent function, making no allowance for interconnections? What area of PCB would be required to realize the equivalent of a 2006-made chip with 150 million transistors?

3.2 Moore's law says that IC complexity doubles approximately every 18 months. How many transistors would have been expected on the largest chips in production in 2006, when this third edition was published? Assume 10^6 transistors per chip in 1987 when the first edition of this book was published. Compare the result with the actual number of transistors on the largest chips made in 2006 (see Figure 3.1).

Power sources and power supplies

4

Objectives

- [ ] To introduce the main sources of electrical energy used in electronic systems, including mains supplies, batteries, and photovoltaic cells.
- [ ] To introduce the concept of a power supply.
- [ ] To discuss the characterization and performance of power supplies.
- [ ] To explain the functions of the main subcircuits found in a power supply.
- [ ] To explain the operation of linear and switching voltage regulators.

All electronic circuits and systems require energy to operate. Energy is required to move electric charge; to produce heat, light, or sound; to produce mechanical movement; and to manipulate information (as in a computer). Energy is a conserved physical quantity: in a closed system energy can be neither created nor destroyed, although it can be converted from one form to another. In electronic engineering, we are usually concerned with electrical energy, although other forms of energy are also important. Heat, for example, is produced in electronic circuits, usually as a by-product of a useful function, and is discussed in Chapter 7. Energy may be stored as chemical energy in a cell or battery. Chemical energy sources are discussed later in this chapter.

Energy is the capacity to do work. The International System of Units (SI) unit of energy is the joule (J).

While the importance of energy should not be forgotten, electronic engineers more frequently use the concept of power. Power can be used to quantify the rate at which heat is produced in a resistor, the mechanical output of a motor, or the rate at which an electronic system takes energy from its energy source.

Power is the rate of conversion, utilization, or transport of energy. The SI unit of power is the watt (W), which is 1 J s^{-1}.

The terms *a.c.* and *d.c.* stand for alternating current and direct current respectively. We customarily talk about a.c. voltage and d.c. voltage, even though technically this is nonsense — what is an "alternating-current voltage?" We can avoid the term *d.c.* by talking about steady voltages and currents, and, in the frequency domain, zero frequency (for example, "A low-pass filter has unity gain at zero frequency").

Let us examine the form in which electronic circuits of various types require their electrical energy supply. If we leave aside a.c. power control systems such as thyristor motor controllers, most circuits operate from one or more d.c. supply rails. Some circuits require only one rail and associated return (usually at 0 V), while others require two rails that are often symmetric with a 0 V return. Logic circuits and microprocessors, for example, usually operate from a single +5 V or +3.3 V rail, or as low as 1.8 V for some modern circuits, while linear circuits such as active filters using operational-amplifiers (op-amps) require perhaps ±12 V rails. Most low-power electronic circuits use voltages below 20 V with respect to earth. Power amplifiers may require supply rails of up to 300 V. Cathode-ray tubes (CRTs) used in older televisions, oscilloscopes, and visual display units required voltages of up to 5 kV or more, although normally at low current.

Apart from a nominal d.c. voltage, what other characteristics of a d.c. supply rail need to be specified? First, the voltage tolerance of the rail must be defined. A nominal +5 V rail supplying low-power Schottky transistor–transistor logic (LS TTL) logic, for example, could be allowed to vary from +4.75 to +5.25 V because this is the recommended operating voltage range for LS TTL circuits. Second, the allowable ripple on the rail must be defined. Ripple is a small a.c. component superimposed on the mean d.c. level and is usually due to the a.c. source from which the d.c. rail has been derived. Note that the ripple waveform is not usually sinusoidal. Ripple is specified by its frequency and its peak-to-peak amplitude.

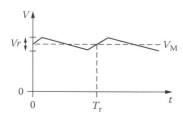

A typical ripple waveform with d.c. component V_M, ripple component Vr, and ripple frequency $1/T_r$.

Energy sources

As designers of electronic apparatus, we are concerned with the form in which energy is to be supplied to or contained within our equipment. The ultimate source of energy used by our design is of interest only in so far as this affects the form, availability, and cost of the energy delivered to our system. We can distinguish two main forms of electrical energy that might be supplied to an electronic system: a.c. or d.c.; and several ways in which an energy source can be built into an electronic apparatus. Let us examine the characteristics of each of these.

A.C. mains

Electronic equipment is often supplied with electrical energy from an external a.c. supply. The most common case is the mains supply derived from an electrical grid. The ultimate source of energy provided by a grid may be coal-fired, oil-fired, nuclear or hydroelectric power stations, diesel or gas-turbine generators, or wind turbines. Grid supplies are characterized (in the developed countries at least) by high reliability and low-cost energy (compared to other sources). A.C. supplies may also be found on board trains, aircraft, and ships where (except in the case of electric trains) the supply will be provided by generators driven from the engines.

Electronic equipment is usually operated from a single-phase a.c. supply.

A.C. supplies are widely used because of the efficiency and ease with which a.c. can be transformed from one voltage to another. Alternating supplies are characterized by their voltage and their frequency. Three common nominal frequencies are used: 50 Hz on the grid systems of Europe, Africa, Asia, and Australia; 60 Hz in North and South America; and 400 Hz on board aircraft. The higher frequency used on aircraft is due to the reduced size and weight of transformers operating at the higher frequency.

R.m.s. stands for root mean square — the square root of the time-averaged value of the square of a waveform:

$$V_{rms} = \sqrt{(\overline{V^2})}$$

Nominal mains voltages vary from one country to another. 230 V a.c. root mean square (r.m.s.) is normal in Europe, although 110 V is often found on building sites and in some factories where the lower voltage is used for safety reasons. In the United States, 120 V a.c. r.m.s. is commonly used. A.C. supplies may vary in voltage, frequency, or both from their nominal values. These variations are due to load variations on the supply — it is common, for example, in the United Kingdom for the frequency to drop by up to 1.5 Hz when a very popular television programme such as a major sporting event ends. This effect is caused by millions of people switching on electric kettles to make cups of tea and therefore creating a sudden increase in electrical load that the power stations cannot instantly meet. The frequency drop is usually corrected within a few seconds as the power stations supply more power. Local voltage drops may also occur, particularly in premises located some distance along a supply cable, as consumers nearer the substation switch on heavy loads. On small electricity grids, voltage and frequency variations may be more marked, and the design of electronic equipment may have to take this into account.

Apart from voltage and frequency fluctuations, a.c. supplies may also be subject to momentary interruption (as, for example, when the local electricity company switches a substation from one circuit to another) and may also carry electrical interference in the form of high-frequency periodic disturbances, switching transients, or harmonics of the grid frequency. This interference may be caused by other electrical or electronic equipment connected to the supply or by natural phenomena such as lightning discharges, and could have a serious effect on the operation of electronic systems that are not designed to cope with it.

Internal energy sources

If a piece of electronic equipment has to be self-contained with its own power source, there are only two options (neglecting portable diesel engines and generators). These are electrochemical cells and photovoltaic cells. Hand-held mobile phones are an example of self-contained equipment powered by electrochemical cells. Electrochemical cells directly convert chemical energy into electrical energy. They can be divided into two types: batteries and fuel cells. Photovoltaic cells directly convert the energy of visible or ultraviolet (UV) light into electrical energy. They are, of course, tremendously important on spacecraft such as communications satellites.

Fuel cells convert chemical energy from reactants supplied externally to the cell into electrical energy. Their best-known application is on board manned spacecraft, such as the NASA Space Shuttle where hydrogen–oxygen fuel cells supply electrical power (and drinking water as a useful by-product). Fuel cells will not be covered further in this book.

Batteries

Batteries are closed electrochemical power sources. They convert chemical energy from reactants incorporated into the device during manufacture to electrical energy. Originally the term *cell* was used in this context, a battery being a collection of cells wired in series or parallel. In modern usage, cells are often referred to as *batteries*.

Two main types of battery exist, known as primary and secondary. Primary batteries can be used once only: when the chemical reactants have been consumed as a result of electrical discharge of the battery, the

---

For a sinusoid, the r.m.s. value is $1/\sqrt{2}$ times the peak value, so the peak value of the European 230 V a.c. r.m.s. mains supply is $230\sqrt{2}$ V or 325 V.

Harmonics of the grid frequency may be caused by thyristor-controlled equipment. For further details, see Bradley (1995).

A very specialized alternative, used for deep space probes, is a nuclear thermoelectric generator that uses the heat from radioactive decay of a mass of plutonium to provide electrical power.

Many batteries contain toxic metals such as lead, cadmium, and mercury, and their safe disposal or recycling should be considered when considering their use. Alternatively, the use of less environmentally damaging batteries should be considered wherever possible. Batteries are exempted from the European Union's Restriction of Hazardous Substances (RoHS) Directive because of the lack of alternatives to lead and cadmium.

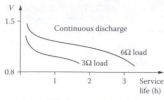

A typical discharge characteristic for two different loads.

device must be discarded. Secondary batteries are based on a reversible chemical reaction: the battery may be recharged by passing electrical current through the device in the opposite direction to the discharge current. Despite their ability to be recharged, secondary batteries have a finite useful life: eventually, recharging fails to store sufficient chemical energy in the battery for useful operation.

Different types of battery have different nominal open-circuit voltages, depending on the electrochemical reaction used in the cells and the number of cells in the battery. The actual voltage provided by a battery falls as the energy stored in the battery is used. The change in voltage is shown on a discharge characteristic for stated conditions of discharge, such as continuous discharge into a specified resistance or intermittent discharge for a specified period per day.

Whether the capacity is stated in Wh or Ah, it is a measure of the energy available from the battery, not its ability to store electric charge (as in a capacitor): the mechanism of energy storage in a battery is chemical, not electrical.

When comparing and selecting batteries, we are usually interested in the amount of energy that the battery can supply before it is fully discharged. This quantity is called the capacity of a battery and can be stated in watt-hours (Wh), which is a unit of energy, or more frequently, in ampere-hours (Ah), which is a unit of electric charge.

Since a battery normally has a fairly constant voltage during discharge, we can calculate the approximate energy content of a battery by multiplying its capacity in Ah by its nominal voltage, remembering that an ampere-hour is 3600 coulombs because there are 3600 seconds in an hour.

Worked Example 4.1

Calculate the energy content of a 2 Ah 12 V battery in (a) watt-hours and (b) joules.

Solution

(a) 2 Ah × 12 V = 24 Wh
(b) 2 Ah × 3600 × 12 V = 86.4 kJ

The concept of the C rate is an approximation, valid only over a limited range of discharge rates.

As in many other areas of electronic engineering, the concept of a normalized quantity is useful when comparing different systems. Battery discharge rates are often expressed in a normalized form known as the "C" rate. A discharge current of 1 C will discharge a battery in 1 hour. A rate of C/5 will discharge a battery in 5 hours, and a rate of 5 C will discharge a battery in 12 minutes. For a 2 Ah battery, the C/5 rate is 400 mA. The C rate can also be used to quantify charge rates for secondary batteries.

The capacity of a battery is not a precise measure of the energy available because the amount of energy that can be extracted from the battery depends upon how it is used. Some types of battery provide more energy in total if they are used intermittently and are allowed to "rest" between discharges, while others are more suited to continuous discharge at a steady rate. When discussing capacity, it is usual to say that the capacity of a battery is dependent on the pattern of discharge, although in reality, it is the amount of energy that can be extracted that is variable.

All batteries will self-discharge to some extent when not in use, because chemical reactions occur within the battery even when no current is being drawn. This limits the performance of secondary batteries and the shelf-life of primary batteries (typically to a few months or years), except for some types of lithium primary battery, which have very long shelf lives.

Worked Example 4.2

A portable VHF radio transceiver consumes 3 W of power at 15 V when transmitting and 0.5 W when receiving only. If the unit is to operate from a secondary battery over an 8-hour shift, what battery capacity is required, assuming the radio is transmitting for a total of 30 minutes during the shift? What has been assumed about the battery capacity?

Solution

Transmitting:

$$\frac{0.5 \text{ hours} \times 3\,\text{W}}{15\text{V}} = 0.1\,\text{Ah}$$

Receiving only:

$$\frac{7.5 \text{ hours} \times 0.5\,\text{W}}{15\text{V}} = 0.25\,\text{Ah}$$

Total energy = 0.35 Ah

A battery of 0.35 Ah capacity is needed, assuming that the intermittent loading will not reduce the battery's capacity and making no allowance for loss of capacity with life.

Main primary battery types

The most widely used primary battery is the Leclanché cell and its variants. Originally, Leclanché cells were wet systems in glass jars, but nowadays the cells are made using chemical reactants in paste form and are referred to as "dry" cells. There are three cell types in common use based on the Leclanché cell. The first of these is known as a zinc–carbon cell from its construction with a carbon rod electrode down the centre of a zinc can that serves as a case and the outer electrode. The paste electrolyte in the cell consists of ammonium chloride and zinc chloride mixed with manganese dioxide. This is the cheapest of the three Leclanché cells and is best suited to applications where current is drawn from the battery intermittently (say, for 1 hour per day). The available capacity of the zinc–carbon cell varies with the pattern of discharge, so that it is not possible to state a single capacity for a cell. The capacity is also reduced at low temperatures, making these cells unsuitable for use in equipment to be used out of doors in freezing conditions. The second variant of the Leclanché cell is known as zinc chloride and has all the ammonium chloride of the zinc–carbon cell replaced by zinc chloride. This gives greater capacity at high current drains and better low-temperature performance, although at greater cost than with the zinc–carbon cell. The third variation on the Leclanché cell is the alkaline–manganese cell, which has a potassium hydroxide (alkaline) electrolyte. These cells

Zinc–carbon cells are now available in a limited range of sizes. They have been replaced by zinc–chloride cells.

are suitable for continuous high-current discharge and have about four times the capacity of a similar sized zinc–carbon cell used in this way. Their low-temperature performance is similar to that of zinc–chloride cells but at higher cost. The shelf life of alkaline–manganese cells is good — they will retain up to 80% of their initial capacity after 4 years in storage at 20°C. All three Leclanché types have a nominal open-circuit voltage of 1.5 V.

Several types of cell are based on lithium. These were originally developed for military applications in, for example, munitions, where their exceptional shelf life of over 10 years was important. Lithium cells have a very wide operating temperature range down to below –20°C and up to over 50°C. Lithium cells are potentially more hazardous than other types of cells, and manufacturer's data sheets and safety advice should be followed carefully.

For miniature equipment such as watches and calculators, there are three primary cell types manufactured as "button" cells. These are the zinc–mercuric oxide cells with an open-circuit voltage of about 1.35 V, the zinc–silver oxide cell with an open-circuit voltage of about 1.6 V, and lithium cells with an open-circuit voltage of 3 V. The first two types have a flat discharge characteristic: the cell voltage remains fairly constant until the battery is almost fully discharged. Zinc–silver oxide batteries are also made in large sizes for military applications such as missiles and torpedoes, where they are used for their high capacity per unit mass despite their high cost.

Table 4.1 summarizes the main types of primary battery.

Main secondary battery types

The main types of secondary battery are the lead–acid type used in road vehicles for starting, lighting, and ignition (SLI) and the nickel–cadmium (NiCd) type used in aircraft and military vehicles, both of which are also available in smaller sizes for powering portable equipment; and the nickel metal hydride (NiMH) type widely used in laptop computers.

The lead–acid cell consists of metallic lead electrodes and sulphuric acid electrolyte. Lead–acid cells have a nominal open-circuit voltage of 2 V. The capacity of lead–acid batteries drops very rapidly below 0°C. Vehicle batteries account for a large proportion of lead–acid battery production. They must support short intense discharge of up to 5 C on engine starting. Automobile batteries are rated at 30 to 100 Ah at 12 V, while commercial vehicle batteries have capacities of up to 600 Ah at 24 V. Larger batteries are used for traction applications such as electric road vehicles and railway locomotives.

For portable equipment, sealed lead–acid batteries are available with capacities from 2 to 30 Ah. These are of lower cost than nickel–cadmium batteries, but are heavier for the same capacity. They have a life in excess of 300 charge–discharge cycles and are maintenance free, needing no topping up of the acid electrolyte. They exhibit a fairly constant voltage during discharge at up to the C/4 rate, and can withstand short high-rate discharge.

Nickel–cadmium batteries are based on cadmium and nickel oxide electrodes with a potassium hydroxide electrolyte. The open-circuit

Table 4.1   Main primary battery types

Type	Nominal open-circuit voltage	Main characteristics
Zinc–carbon (Leclanché)	1.5 V	Low cost. Best used intermittently. Poor performance at low temperature.
Zinc–chloride (Leclanché)	1.5 V	Improved capacity at high current drain compared to zinc–carbon, and better low-temperature performance.
Alkaline–manganese	1.5 V	Suited to high-current continuous discharge. Long shelf life. Similar low-temperature performance to zinc chloride but at higher cost.
Lithium–thionyl chloride and	3.5 V	High cost. Long shelf life. Wide operating temperature range down to –20°C or less and up to +50°C or more.
Lithium–manganese dioxide	3 V	
Zinc–mercuric oxide	1.35 V	Available as "button" cells for miniature equipment such as watches. Flat discharge characteristic. Good shelf life.
Zinc–silver oxide	1.6 V	High cost. Flat discharge characteristic. High capacity per unit mass in large sizes. Military applications.

voltage is about 1.2 V. Nickel–cadmium batteries are more expensive than lead–acid batteries and are the most important alkaline secondary type. Unlike lead–acid cells, they can work well at temperatures down to less than –30°C. They have a flat discharge characteristic and can accept continuous overcharging at a low charge current. (This is known as "trickle" charging.)

Nickel metal hydride batteries are similar to nickel–cadmium ones but are less environmentally hazardous. Their capacity is greater than that of NiCd cells, and they are widely used in hybrid vehicles (for example, the Toyota Prius) and in portable electronic devices such as phones and laptop computers. They have a higher internal resistance than NiCd cells and therefore are less suited to applications with high current demand.

For further reading on batteries, see Vincent and Scosati (2000).

Lithium ion batteries are also widely used for portable devices, particularly laptop computers. Their light weight and slow self-discharge are useful, but they lose capacity due to ageing, with or without use.

Photovoltaic cells

Photovoltaic cells convert the energy of visible or ultraviolet light into electrical energy. They are used on board communications satellites to supply electrical power where no other energy source other than a radioisotope generator or small nuclear reactor could supply the power needed over the many years of the satellite's life. Solar cells may also be used on Earth, for example to power communications relay stations located in remote regions far from electrical grids, or roadside illuminated signs. In this case, electrochemical cells are needed to store energy for use during the night when direct solar power is not available. On a smaller scale, some electronic watches and calculators are powered by photovoltaic cells with secondary electrochemical cells providing energy storage to power the device while in darkness or low light.

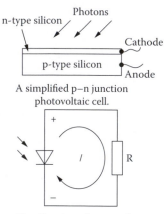

Other cells are based on cadmium sulphide, selenium, and gallium arsenide. All include a p–n or semiconductor–metal junction.

A simplified p–n junction photovoltaic cell.

The direction of current flow from a photovoltaic cell.

The most important photovoltaic cells are silicon junction diodes of large area with a thin n-type region on the exposed face.

A p–n junction with no externally applied bias develops a space charge region on either side of the junction as mobile charge carriers diffuse across the junction under the influence of concentration gradients. In equilibrium, there is no net charge transport across the junction because of the presence of an electrostatic potential.

If electromagnetic radiation of suitable wavelength now illuminates the junction, electron-hole pairs can be created by photon absorption. This process can be thought of as the ionization of a silicon atom, creating a free electron and a positively charged silicon ion. The silicon ion can attract an electron from a neighbouring atom, so that the positive charge (a hole) is mobile. The free electron and the free hole are, however, influenced by the electrostatic potential difference across the junction: they move in opposite directions, the electron towards the cathode and the hole towards the anode. If the diode is connected to an external circuit, current can flow and supply power to the external circuit. The direction of this current flow is that of a reverse current through the diode: the positive potential is developed at the anode. This means that in a circuit where a photovoltaic cell charges a secondary battery, a blocking diode must be connected in series with the cell to prevent the battery from driving forward current through the cell during darkness. The blocking diode must have a low forward-voltage drop, and often a Schottky barrier (semiconductor–metal) diode is used. The electromotive force (e.m.f.) generated by a silicon photovoltaic cell is around 0.5 V, so that practical circuits using secondary batteries have several photovoltaic cells in series to raise the e.m.f. to a practical level. Solar cells may also be made from amorphous silicon, and although cheaper, they are less efficient.

The arrow in the diode (and junction transistor) symbol represents the direction of forward current flow when the diode is driven by an external circuit. Conventional current, by historical accident, flows in the opposite direction to electrons.

Power supplies

The abbreviation PSU for *power-supply unit* is often used.

So far in this chapter, we have looked at energy sources and the forms in which energy may be used by electronic circuits. We can now look at the way in which electrical energy can be converted into the required form. In many electronic systems, this conversion is performed by a subsystem called a power supply. Even in battery-operated equipment where the energy source provides energy in almost the required form, there may still be some form of power supply.

A wide range of power supplies are available commercially, usually from manufacturers specializing in this field. Many larger electronic systems use commercial power supply units bought off the shelf. In other cases, however, a custom design may be needed, because for example a special physical form, low weight, or high reliability is required. For high-volume mass production, a custom power supply, as with any other component or subsystem, may be cheaper than any commercial off-the-shelf design because of a better match between the power-supply design and the system requirement. Often such custom design is contracted out to a power-supply firm with special expertise.

We can now look at the main functions of a power supply. The most obvious and common function of a power supply is to convert electrical energy at the source voltage to some other voltage, higher or lower than the source voltage, and with or without a change from a.c. to d.c. or vice versa. A computer, for example, might be designed to operate from an a.c. mains supply and yet contain circuitry operating at 5 V or 3.3 V d.c. A power supply would be needed, therefore, to reduce the voltage and convert the energy to d.c. On board a communications satellite, d.c. will be available from batteries charged from a solar panel. For economy of space and weight, only one voltage will be available from the batteries. Electronic circuits and systems such as microwave amplifiers, attitude controllers, and computers, however, may need a variety of voltages both lower and higher than the battery voltage. D.C.–d.c. converters can produce these voltages at high efficiency without wasting valuable energy as useless heat. Inverters generate a.c. from a d.c. input. One common application is on board small boats to generate 120/230 V a.c. 50/60 Hz from a 12 V battery, allowing low-powered domestic mains-operated equipment such as radios and shavers to be used.

Voltage conversion, whether to a higher or lower voltage, is possible in practical terms only in an a.c. circuit, using a transformer. D.c.–d.c. power supplies (or converters) are in reality, then, d.c. to a.c. to d.c. power supplies.

Power supplies operating from an a.c. source also have to provide energy storage during the parts of the source cycle where little or no energy is available from the supply. It may also be necessary to store energy to supply the output current during momentary loss of the a.c. supply, such as happens when a substation is switched. Energy storage is usually in the form of electric charge in the power supply's reservoir capacitors and is usually practical for only a limited time in most systems. Longer-term energy storage requires the use of batteries. A power supply with batteries is an example of an uninterruptible power supply (UPS). A UPS produces constant output even during breaks in the a.c. mains supply. Laptop computers work in this way, since loss of the mains supply does not interrupt the supply to the computer — this comes from the battery.

A change in voltage is always accompanied by a compensating change in current: the power output of a power supply is always less than the power input.

Voltage increase at low current can be achieved without a transformer using a diode multiplier such as the Cockcroft-Walton multiplier. These circuits do, however, require a switching action.

Larger uninterruptible power supplies may use diesel generators as well as, or instead of, batteries.

Exercise 4.1

A telemetry transmitter uses 4 W of power at 5 V d.c., derived from a mains supply. It is to be capable of continued operation despite momentary interruptions in the mains supply lasting up to 125 ms. Show that a capacitor of about 1 F is needed if connected across the 5 V rail, assuming that the rail voltage must not drop below 4.9 V during the supply interruption.

The third main function of a power supply is maintenance of a constant output irrespective of varying *load*. This is known as regulation or stabilization, and a power supply having this feature is called a regulated or stabilized power supply. Not all power supplies have this feature as some applications can tolerate variations in voltage or current. A d.c. supply for low-voltage filament lamps and relays is an example of such an application.

Power-supply performance can be represented graphically as a characteristic, which shows output voltage as a function of output current. Ideally, the output voltage would be independent of output current, but in practice some drop in voltage is unavoidable and is quantified by a parameter known as load regulation. The definition of load regulation varies somewhat, but a fairly common definition is given below and in Figure 4.1. V_{10} is the output voltage at 10% of the full load current, and V_{100} is the output voltage at 100% of the full load current. Load regulation is then defined as

$$\text{Load regulation} = \frac{(V_{10} - V_{100})}{V_{10}} \times 100\% \qquad (4.1)$$

Output voltage variation might also be due to variation in the source voltage. The a.c. mains voltage, for example, can vary, particularly for consumers at the end of a cable some distance from a substation. The terminal voltage of a battery falls as the battery becomes discharged. It is possible to define a regulation factor to quantify the performance of a

An example of a load that would demand varying power is a power amplifier driving a loudspeaker with speech.

Varying definitions of such practical parameters are quite common — be cautious when using such parameters.

Load regulation is normally expressed as a percentage. An alternative definition might use the full-load and no-load voltages, rather than the full load and 10% of full-load voltages.

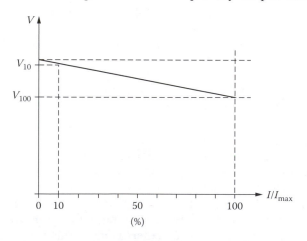

Figure 4.1   Output characteristic of a voltage-regulated power supply showing quantities used in defining load regulation.

power supply under variations of input voltage, but this can be done in many ways and it will not be discussed further here.

A final factor that can influence the regulation of a power supply is ambient temperature. Normally, the variation in output voltage with temperature is small and roughly linear over the operating temperature range of the power supply, so that the normal idea of a temperature coefficient can be used to quantify this aspect of the power supply's performance.

Apart from the precisely controlled voltage, there is another benefit to be had from a voltage-regulated power supply. This is a low a.c. or dynamic impedance and is very important in the operation of linear circuits. In analysing, say, a common-emitter amplifier circuit, we assume the power supply to be of negligible impedance in deriving our small-signal model. We must not forget that this assumption may not always be valid.

Overload protection

We can now turn our attention to the part of the output characteristic beyond the full-load current. What happens when the full-rated current of a regulated power supply is exceeded? Figure 4.2 shows some of the possibilities. If no special provision is made in the design of a power

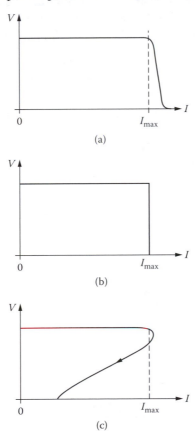

Figure 4.2   Output characteristics of voltage-regulated power supplies: (a) with voltage reduction beyond full-load current, (b) with crossover to constant current at full-load current, and (c) with "foldback" limiting.

Temperature coefficients are discussed in the next chapter in the context of passive components.

63

Current limiting, as shown in Figure 4.2b, is a popular option for bench power supplies used in laboratories. Often the limiting current is adjustable so that limiting can be used to protect the circuit being tested.

supply, damage or destruction of some components may occur due to overheating. The power supply may include a fuse that will rupture and disconnect the power supply from its energy source. This possibility is not shown in the figure. Current limiting is the next most likely possibility — the power supply will be designed so that beyond full-rated load current, the output voltage will be reduced. This may be done by limiting current as in Figure 4.2b. A third, more elaborate possibility is known as foldback limiting and is illustrated in Figure 4.2c. Here, once the full-rated load current has been exceeded, the power-supply output voltage and current are reduced, bringing both down independently of the load.

We have considered protection of the power supply itself from damage due to excessive current being drawn by the load. Foldback limiting also protects the load to some extent. One other eventuality that must be considered, however, is overvoltage. Suppose that a regulated power supply supplies current to a large system containing many expensive integrated circuits (such as a computer). If the power supply becomes faulty and the output voltage becomes too high, a large amount of expensive circuitry could be damaged beyond repair if the voltage exceeded the absolute maximum ratings of the integrated circuits. To guard against this prospect, overvoltage protection is normally included in the power-supply design. Figure 4.3 shows a common solution known as a crowbar circuit. The essence of this circuit is to create a short circuit (or at least a very low-resistance path) across the power-supply terminals as soon as an overvoltage is detected (within microseconds). When the overvoltage-sensing circuit detects that the regulated voltage has risen beyond its normal limit, a triggering signal is generated, switching the SCR or thyristor into its conducting state. R_A limits the maximum current through the SCR and is typically a fraction of an ohm. Once triggered, the SCR effectively short circuits the regulated supply and prevents damage to the load. The SCR continues to conduct until current is switched off.

Overvoltages may also occur at switch-on and switch-off unless the power-supply design deliberately includes proper control such as a defined "power-up" sequence.

The characteristics of thyristors or SCRs are discussed by Bradley (1995).

We have now examined the main functions and characteristics of power supplies and can turn our attention to the circuits used within power supplies. Figure 4.4 is a block diagram of a typical mains-operated power supply. The transformer changes the input a.c. voltage to a higher or lower a.c. voltage, which is then converted to d.c. by the

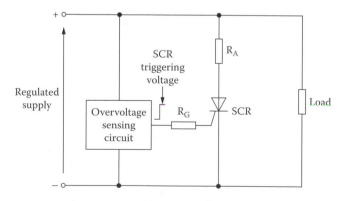

Figure 4.3   A crowbar overvoltage protection circuit.

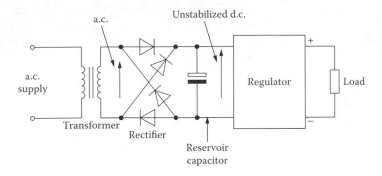

Figure 4.4 Block arrangement of a typical mains-operated d.c. power supply.

full-wave bridge rectifier and reservoir. At this stage, before the regulator, the d.c. voltage is unstabilized (it will vary as the load current varies) and may also have greater ripple than the regulated voltage. The regulator stabilizes the load voltage and may also include overload protection circuits. Practical power-supply designs may not show such a clear distinction among the functional blocks. Let us now consider each functional block in turn.

Transformers

Transformers are widely available commercially in a range of physical sizes, power ratings, and electrical configurations, and are designed for operation at a specific frequency (usually 50, 60, or 400 Hz). The theory of transformer operation is covered elsewhere, and will only be discussed briefly here. Transformers consist of one or more electrical windings of low d.c. resistance, wound onto a core of magnetic material (commonly iron). A changing current in one winding induces a changing magnetic field in the core, which links the same or another winding and induces a changing e.m.f. in that winding. Two main types of transformer are shown in Figure 4.5. The autotransformer (Figure 4.5a) has one winding with intermediate tappings, while the double-wound transformer has separate primary and secondary windings. The double-wound transformer is almost universally used in electronic power supplies, because of the increased safety obtained by having electrical isolation between the windings. The ratio of turns on the secondary winding to the number of turns on the primary winding determines the ratio of secondary voltage to primary voltage. The amount of power that can be drawn from a secondary winding is always less than the power supplied to the primary winding, because some energy is lost as heat in the windings due to their resistance (copper loss) and as heat due to circulating electric currents (eddy currents) in the core (iron loss). Iron losses can be reduced by constructing the core from thin flat sheets of iron, or laminations, stacked together and insulated from each other with a layer of varnish, or by constructing the core from ferrite that is nonconductive. Copper losses can be reduced by increasing the cross-sectional area of the wire used for the windings. Copper is an expensive metal, however, so the transformer designer must trade off copper losses against the cost of the transformer. For small transformers such as are

See for example Senturia and Wedlock (1993).

The primary winding is the energy-input side of the transformer. The secondary winding is the energy-output side of the transformer.

Ferrites are discussed in Chapter 5.

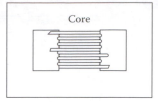

"Shell" type transformer with all windings on centre-limb.

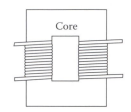

"Core" type transformer with primary and secondary windings on separate limbs.

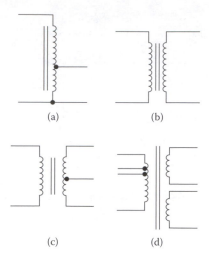

(a)   (b)

(c)   (d)

Figure 4.5   Types of transformer: (a) autotransformer, (b) double-wound transformer, (c) double-wound transformer with centre-tapped secondary winding, and (d) double-wound transformer with multitapped primary winding and separate secondary windings.

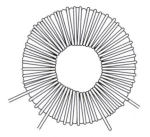

Toroidal transformer.

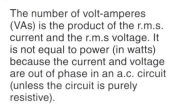

The number of volt-amperes (VAs) is the product of the r.m.s. current and the r.m.s voltage. It is not equal to power (in watts) because the current and voltage are out of phase in an a.c. circuit (unless the circuit is purely resistive).

used in electronics, cost is likely to be more important than a small energy loss.

There are three common constructional designs for double-wound transformers shown diagrammatically in the margin. The shell type is the most common for small transformers and has all the windings wound on a common centre limb of the core. Small core-type transformers are less common and have primary and secondary windings located on separate limbs of the core. The chief advantage of this arrangement is reduced capacitive coupling between the windings. The third transformer shown consists of a toroidal (doughnut-shaped) core and is a compact design popular for low-profile equipment. It also has reduced flux leakage compared to other designs, a factor that can be important in audio-amplifiers and low-frequency instruments because of their tendency to pick up and amplify mains frequency signals. When mounting toroidal transformers, it is most important not to create an electric circuit through the centre of the toroid. This would constitute a short-circuited secondary and would cause overheating and possibly destruction of the transformer. Mountings should either use nonconducting fasteners (such as nylon screws) or be electrically insulated at one or both ends.

The isolation between the primary and secondary windings of a transformer can be less than perfect at higher frequencies because of capacitive coupling between the primary and secondary windings. This coupling may result in transmission of electrical interference (unwanted noise and transients) into the secondary circuit. To reduce the coupling effect, some transformers are fitted with an electrostatic screen between the primary and secondary windings, brought out to an external terminal that is usually grounded.

Transformer ratings are usually expressed in VA (volt–amperes), rather than watts, or by stating the maximum r.m.s. current rating of each winding. Transformer secondaries should be fused, or otherwise

66

protected against damage by a short-circuit load, because a secondary winding can be supplying heavy current under fault conditions without the primary current exceeding the rating of the primary winding. It is therefore not sufficient to rely on a fuse in the primary circuit to protect the secondary.

Rectification

Rectification is the conversion of a.c. to pulsed d.c. Several rectifier circuits are shown in Figure 4.6. The most important rectifying component in modern use is the semiconductor diode. Power diodes may sometimes be referred to as rectifiers to distinguish them from signal diodes. Full-wave rectification is universally used in electronics and produces pulsed d.c. at twice the supply frequency. This can be achieved with two diodes if a centre-tapped transformer is used, or with four diodes connected in a bridge configuration. The bridge configuration is the most widely used nowadays, requiring a secondary winding without a centre tap on the transformer. Four diodes arranged in a bridge with four terminals or leads are commonly available and known as bridge rectifiers.

The centre-tapped full-wave circuit was widely used in the days of thermionic valve electronics, the two diodes being implemented within one valve envelope (a double diode). This is a good example of how cost influences design: at one time extra copper (in the form of a centre-tapped secondary) was cheaper than extra diodes (to form a full-wave bridge).

Rectifier or power diodes have separate ratings for average forward current and surge current. As will be seen in the next section, the inclusion of reservoir capacitors in power-supply circuits can cause large surge currents to flow when the diodes start to conduct in each cycle. Power diodes also have greater forward-voltage drop than small-signal diodes

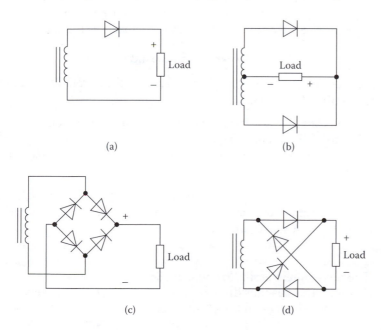

Figure 4.6   Rectifier circuits: (a) half-wave, (b) full-wave, (c) full-wave bridge, and (d) alternative representation of the full-wave bridge.

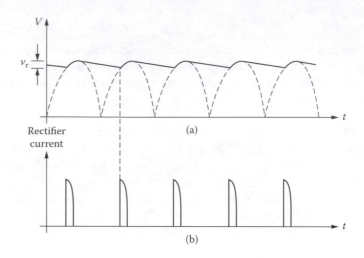

Figure 4.7 (a) Output waveform of a full-wave rectifier circuit with reservoir capacitor, and (b) rectifier current.

so that power dissipation in a rectifier diode can be significant. Lastly, the maximum reverse voltage or peak inverse voltage (PIV) rating of rectifiers must be adequate both for the normal reverse voltages present in the circuit and for any abnormal transient voltages on the supply.

Reservoir capacitors

For the majority of electronic applications, the pulsed d.c. output from a rectifier is unsuitable directly. In some applications, a reservoir capacitor or smoothing capacitor connected between the supply rails is sufficient.

Figure 4.7 shows the effect of a reservoir capacitor on the output of a full-wave rectifier. The ripple voltage, v_r, is determined by the value of the reservoir capacitor, the load current, and the supply frequency (which determines the time over which the capacitor discharges).

Worked Example 4.3

A reservoir capacitor is to be connected across the output of a full-wave bridge rectifier. Calculate:

(a) The phase angle at which the rectifiers start to conduct.
(b) The value of capacitor required.
(c) The peak rectifier current, neglecting any series resistance in the circuit.

The peak value of the rectified waveform is 6.5 V; the supply frequency is 50 Hz. The ripple must not exceed 50 mV peak-to-peak, and the load current is 100 mA maximum.

Solution

(a) The phase angle is $\sin^{-1}(6.45/6.5) \approx 83°$.
(b) Assume the capacitor discharges linearly between the peak of one cycle and the angle in (a):

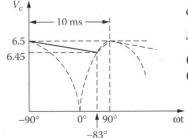

$$C = \frac{\Delta q}{\Delta V} \approx \frac{100\,\text{mA} \times 10\,\text{ms}}{50\,\text{mV}} \approx 20\,\text{mF}$$

(c) The capacitor voltage V_C is given by:

$V_C = 6.5 \sin \omega t$, where $\omega = 2\pi \times 50$ rad/s, $83° \leqslant \omega t \leqslant 90°$.

The rate of change of V_C is $\dfrac{dV_C}{dt} = 6.5\omega \cos \omega t$, and the greatest rate

of change occurs when $\omega t = 83°$.

$$\left.\frac{dV_C}{dt}\right|\max = 6.5(2\pi) \times 50 \times \cos 83° \approx 250\ \mathrm{V}s^{-1}$$

The peak rectifier current is given by:

$$\left.I_R\right|\max = C\frac{dV_C}{dt} + 100\,\mathrm{mA} = 20\,\mathrm{mF} \times 250\,\mathrm{V}s^{-1} + 100\ \mathrm{mA} \approx 5.1\,\mathrm{A}$$

Voltage references

All regulated power supplies require a voltage reference: a device or circuit that can maintain a constant voltage between its terminals independently or nearly independently of variations in current or ambient temperature. Before considering the various types of regulator, let us consider the most commonly used voltage reference component, the Zener diode.

Figure 4.8 shows the *V-I* characteristic of a typical Zener diode. The characteristic shows only the reverse-biased region because Zener diodes are operated in reverse breakdown. Their forward characteristic is of no interest. The main feature of the characteristic to note is that, for voltages greater than the breakdown voltage, V_Z, a small increase in voltage produces a large increase in current. Looked at another way, the voltage across the diode is nominally constant over a wide range of currents.

The *V–I* characteristic in the breakdown region is not precisely parallel to the current axis (a change in current does produce a small change in voltage). This variation is expressed as the diode's slope resistance, R_Z (in ohms), and is the ratio of incremental voltage change to incremental current change at a specified current. R_Z is not constant, but

There are low-voltage reference diodes (which are actually integrated circuits) that use the band-gap energy of silicon to provide a reference voltage. These are known as band-gap references.

All junction diodes exhibit reverse breakdown, but usually at much higher voltages. Zener diodes are specially fabricated to break down at a precise voltage and to withstand continuous power dissipation in the breakdown region. The name *Zener diode* is actually a misnomer for diodes with breakdown voltages above about 5 V. The Zener effect occurs when electrons within the space-charge region of the diode are dislodged from their atoms by the electric field intensity. The dominant process in most diode breakdown is avalanche multiplication, where the energy of dislodged electrons is sufficient to dislodge further electrons from their bonds.

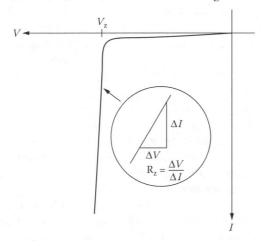

Figure 4.8   *V-I* characteristic of a Zener voltage-reference diode.

varies with current. Typically, Zener diodes have slope resistances from a few ohms to a few tens of ohms. Breakdown voltages also vary with temperature, the variation being expressed in the usual way as a temperature coefficient.

Worked Example 4.4

A 5.1 V type BZX79 500 mW Zener diode has a slope resistance of 60 Ω maximum at 5 mA. What would be the allowable range of currents if the diode voltage was not to vary by more than ±0.1 V?

Solution

$$\frac{\Delta V}{\Delta I} = R_Z \qquad R_Z = 60\,\Omega \text{ and } \Delta V = \pm 0.1\,\text{V}$$

thus

$$\Delta I = \pm \frac{0.1\,V}{60\,\Omega} \simeq \pm 1.7\,\text{mA}$$

The allowable range of currents is thus 3.3 to 6.7 mA.

Zener diodes are available in power ratings from 400 mW to over 20 W. Note that a Zener must be operated at a suitable current (the operating point must be beyond the "knee" point of the *V-I* characteristic). Manufacturer's data usually state the slope resistance at a specified current, and generally, the diode should be operated near to the stated current.

Linear regulators

Linear voltage regulators operate by dropping an unregulated voltage through a dissipative element (either a resistor or a transistor), controlling the voltage drop so as to maintain a constant output voltage. Two possible arrangements of linear regulator are shown in Figure 4.9. The load represents the electronic circuits to be supplied with a constant voltage. It contains active circuits, and its impedance varies with time. In the series regulator, a regulating element is placed in series with the load. The voltage drop across the regulating element is varied as the load current or the unregulated voltage varies in order to maintain a constant voltage across the load. In the shunt regulator, a resistor is placed in series with the load and a regulating element is placed in parallel with the load. The current drawn by the regulating element is varied in order to alter the voltage drop across the resistor and keep the load voltage constant. Both series and shunt regulator circuits are used in practical designs, as we shall see below.

The simplest arrangement of linear voltage regulator is a shunt circuit and is shown in Figure 4.10. The regulated voltage V_R is obtained directly across the Zener diode D_Z. When the load current I_L or the unregulated voltage V_U changes, the Zener diode current I_Z changes to compensate.

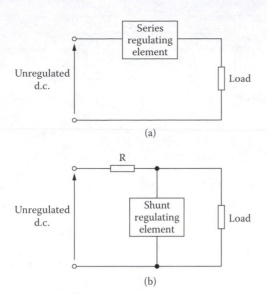

Figure 4.9 Two arrangements of linear voltage regulators: (a) series, and (b) shunt.

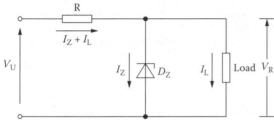

Figure 4.10   A simple linear shunt regulator.

Worked Example 4.5

Some complementary metal-oxide semiconductor (CMOS) logic circuits operating at +3.3 V are to be included in a system and powered from a simple Zener shunt regulator circuit of the type shown in Figure 4.10. There is a regulated +5 V rail available. The designer decides to use a 300 mW 3.3 V Zener and finds from a data sheet that a typical recommended operating current is 5 mA. Since CMOS logic draws negligible current (micro-amperes) when quiescent, voltage regulation must be maintained down to zero load current. Select a value for the resistor, and calculate the maximum allowable load current. What factors have been neglected in the calculation?

Solution

Maximum Zener current $I_{Z,\text{max}} = \dfrac{225\,\text{mW}}{3.3\,\text{V}} \approx 68\,\text{mA}$

Assume load current $I_L = 0$, then $R = \dfrac{5 - 3.3}{0.068} = 25\,\Omega$

$I_Z + I_L$ is constant at 68 mA. I_Z must not fall below 5 mA, therefore $I_{L,\text{max}} = 63$ mA.

The self-heating, slope resistance, and temperature coefficients of the Zener diode have been neglected.

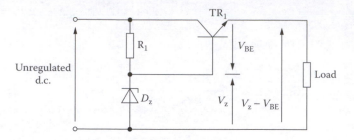

Figure 4.11   A simple linear series regulator.

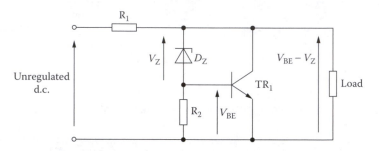

Figure 4.12   A transistor shunt regulator.

The simple Zener shunt circuit has several shortcomings, among which are that the Zener diode current and therefore the power dissipated in the diode vary with load current and unregulated voltage and that there is no compensation for the temperature coefficient of the diode. The simple series regulator described next partly overcomes the first of these problems. Temperature compensation of Zener diodes can be achieved by adding other devices with a similar temperature coefficient of opposite sign in series with the Zener.

Figure 4.11 shows a simple linear series regulator design using a bipolar transistor as the dissipative element. The transistor is in series with the load. The voltage at the emitter of the regulating transistor is constant at V_{BE} less than the reference voltage across the Zener diode. R_1 supplies operating current for the Zener diode and base current for the transistor. Variations on this simple circuit exist to overcome problems with temperature stability of the Zener voltage and V_{BE} drop. Protection against short circuit of the load terminals may be needed, since under these conditions the transistor passes heavy current with the full unregulated voltage across the transistor.

Figure 4.12 shows a simple transistor-based shunt regulator. In this design, the dissipative element is the resistor R_1 (although some power is also dissipated in the shunt transistor). The shunt regulator has the important advantage of being inherently protected against a short-circuit load, because under these conditions the transistor passes no current. Another useful feature of this circuit is that it provides a path for reverse current from the load and can actually absorb power from the load. This is an advantage when the load is a d.c. motor.

Integrated-circuit voltage regulators are available for commonly used voltages such as 3.3V, 5 V, and 12 V and with adjustable voltage outputs. They are available in both positive and negative polarities and include a voltage reference, regulating element, and control circuits within the package. They usually include current limiting and temperature compensation. Low-power types may be packaged in dual-in-line form or surface mount packages. Higher-power devices are packaged in the same way as power transistors and require heat sinking.

Switching regulators

Switching regulators operate at higher efficiency than linear types by avoiding power wastage in a series- or shunt-regulating device. They are also smaller and lighter for a given power output than linear regulators. Their disadvantage is that, because of the switching action, output ripple is usually higher than with linear regulators. Also, because of their greater complexity and higher component count, switching regulators are usually slightly less reliable than linear regulators. (Even so, some switching regulators have mean time between failures [MTBFs] of over 200,000 hours, or more than 20 years.)

MTBF, or *mean time between failures*, is discussed in Chapter 9.

Switching regulators operate by chopping (switching on and off) the unregulated voltage, matching the demanded power with supplied power. A smoothing circuit produces continuous d.c. from the chopped waveform. The smoothed output is sensed and fed back to control the chopping frequency or pulse width.

Figure 4.13 shows a simple example to show the principle (in reality, circuits are more complex). The circuit operates as follows. The control waveform generator produces a control signal consisting of rectangular pulses that causes TR_1 to chop the input voltage. When TR_1 is conducting, current flows to the load through the inductor L_1. Capacitor C_1 is charged to the output voltage of the regulator. When TR_1 switches off, current continues to flow through L_1. The flywheel diode D_1 is required to hold down the voltage at the collector of TR_1 and to provide a path for the current while the transistor is off. During the time that the transistor is off, the load current is being supplied from the stored energy in the inductor and capacitor. The output voltage is thus a mean

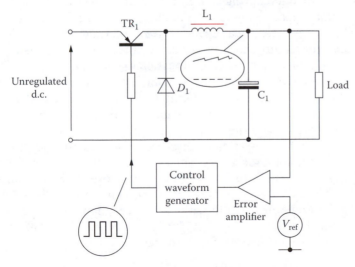

Figure 4.13   A switching voltage regulator.

d.c. level with superimposed ripple. The mean level is regulated by comparing it with a voltage reference and generating an error signal to control the pulse generator. The output voltage may be varied by changing the frequency of the pulses, keeping their width constant, or by varying the pulse width, keeping the frequency constant. Either way, the mark-to-space ratio or duty cycle of the pulse generator output varies and controls the output voltage of the regulator.

Typical operating frequencies of switching power supplies are above 20 kHz, both to avoid generation of audible frequency acoustic noise and because inductors for higher-frequency operation can be made smaller and lighter than those for lower frequencies.

Summary

This chapter has looked at the main energy sources used to power electronic systems, starting with the characteristics of a.c. mains supplies and then examining energy sources such as electrochemical and photovoltaic cells, which can be included within self-contained systems.

The idea of a power supply as a subsystem within an electronic system has been introduced, and the performance characterization of power supplies has been discussed. The main elements of a power supply were then described, including transformers, rectifier circuits, and reservoir capacitors. Voltage regulation was then introduced, and the two types of regulator (linear and switching) were outlined.

The chapter has thus given a broad introduction to the subject of powering electronic equipment.

Problems

4.1 A D-size nickel–cadmium cell has a nominal 4 Ah capacity and a nominal open-circuit voltage of 1.25 V. The manufacturers recommend a 12-hour charge at 500 mA. What will be the average heat dissipation during charging assuming that the cell absorbs its nominal capacity in the form of chemical energy?

4.2 The energy stored in a capacitor is $\frac{1}{2} CV^2$. Calculate the value of a capacitor required to store the same amount of energy as a 30 Ah 12 V car battery if the capacitor is charged to 12 *V*.

4.3 Why would a primary battery manufacturer advise users to store batteries in a cool place and keep them out of direct sunlight?

4.4 Recalculate (a) the capacitor value and (b) the peak rectifier current of Worked Example 4.3 for a supply frequency of 400 Hz.

4.5 What is (a) the maximum discharge rate and (b) the minimum charge rate for a secondary battery used to power a miner's helmet lamp if the lamp is to operate continuously over an 8-hour shift and then be ready for use at the start of the same shift on the following day? Assume that two thirds of the energy input to the battery is absorbed as chemical energy during charging. (Hint: The answers are expressed independently of the battery capacity.)

4.6 Calculate the efficiency of the circuit designed in Worked Example 4.5 at (a) maximum load current and (b) a load current of 5 mA. Efficiency is the ratio of power delivered to the load to power drawn from the source (in this case, the +5 V rail).

4.7 For the application given in Worked Example 4.5, an engineer decides to try the circuit of Figure 4.11, and sets the Zener diode current at 5 mA by suitable choice of R_1. Calculate the efficiency for the same load currents as Problem 4.6. Neglect the transistor base current.

Passive electronic components

5

Objectives

☐ To emphasize the differences between real components and ideal circuit elements.
☐ To introduce the properties and characteristics of real passive components.
☐ To survey the main types of passive electronic component and to discuss their fabrication.

Passive component characteristics

A fully detailed data sheet for an apparently straightforward component such as a capacitor contains information on a great many aspects of its behaviour. In any particular application, some of the component parameters will be of the utmost importance, while others will be of little consequence. In some circuits, the ultimate performance that can be achieved is limited by component behaviour. Figure 5.1 shows such a circuit, a single-slope analogue-to-digital converter, or ADC. In this type of ADC, a ramp generator or integrator circuit produces a ramp waveform starting at a voltage slightly below 0 V. Two comparators are used, one to compare the ramp waveform with 0 V and one to compare the analogue input voltage with the ramp waveform. As the ramp waveform increases from below 0 V to greater than the analogue input, the two comparators switch one after the other, and the time interval between the two comparators switching is accurately proportional to the difference between the analogue input voltage and 0 V. The comparator outputs are used to start and stop a counter driven from a clock circuit. The final digital value in the counter is proportional to the time interval between the switching of the two comparators and therefore to the analogue input voltage. Figure 5.1b shows the details of the ramp generator circuit, which integrates a constant reference voltage to produce a ramp. A field-effect transistor (FET) switch is needed to reset the integrator at the end of each cycle by discharging the capacitor. (The control signal for the FET has been omitted from Figure 5.1b for the sake of clarity.) The overall linearity of the ADC depends critically on the quality of the ramp waveform: any significant deviation from an ideal ramp, as in Figure 5.1c, will cause linearity errors in the ADC output. One possible cause of ramp nonlinearity is the integrator capacitor: leakage current through the capacitor dielectric causes the ramp waveform to droop. A further problem is that capacitor dielectrics can absorb charge, a phenomenon that is discussed later in this chapter. The choice of capacitor for this circuit is thus very important as the linearity of the ADC depends on the capacitor.

The single-slope analogue-to-digital converter (ADC) is a **classic** circuit. Typical ADC circuits used in practice are more elaborate.

Linearity is a very important concept in electronic engineering. A linear system obeys the principle of superposition, such that if input x_1 causes output y_1 and input x_2 causes output y_2, then an input of $x_1 + x_2$ will cause an output of $y_1 + y_2$. In the case of an ADC, linearity means that the value of the digital output is accurately proportional to the analogue input.

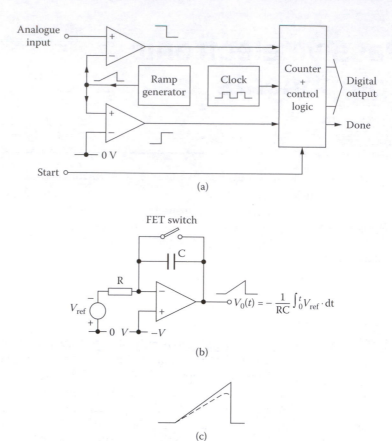

$$V_0(t) = -\frac{1}{RC}\int_0^t V_{ref} \cdot dt$$

Figure 5.1 A single-slope analogue-to-digital converter: (a) block diagram, (b) ramp generator circuit, and (c) ramp nonlinearity.

Exercise 5.1

Explain why two comparators are used in the single-slope ADC discussed above, and why the counter is not started at the start of the ramp.

An understanding of component parameters is essential to the circuit designer who needs to compare different types of component for a particular application. A good circuit designer comes to know the type of component that will be needed for a particular purpose and understands why that type of component will do the job.

This chapter starts by looking at some general aspects of passive components, and then moves on to discuss specific types of component.

Tolerance

In engineering, no manufactured value or dimension can ever be exact. Engineers express the closeness of a value or dimension to the desired value by a **tolerance**. Smaller tolerances are more difficult to achieve, and components with small tolerances are therefore more expensive. The acceptable range of a value can be specified in several ways. One is to state the upper and lower limits of the range within which the value must lie. This method is sometimes used in mechanical dimensioning, but is rarely used for component values in electronic engineering, where the normal practice is to state a **nominal** value with a **tolerance**. Often, but not necessarily, the nominal value lies in the middle of the acceptable range and the tolerance is quoted as a percentage of that nominal value. A 100 Ω ± 5% resistor, for example, could have any value from 95 to 105 Ω. For some components, the tolerances above and below the nominal value are unequal, implying perhaps an uneven probability distribution resulting from the fabrication process.

Electrolytic capacitors commonly have capacitance tolerances of ±20% but some are specified asymmetrically, for example as −10% to +30%.

The tolerances described so far have been **relative**: they represent a fractional deviation from the nominal value. For some components such as capacitors in the 1 to 10 pF range, the tolerance may be expressed as an **absolute** value thus: 3 pF ± 0.5 pF.

Preferred values

Clearly it is not possible for a resistor manufacturer to produce economically every possible value of resistance, even to a 1% tolerance. Accordingly, resistors (and capacitors) are manufactured to limited ranges of **preferred values**. The most common range is known as E12 and contains values of 10, 12, 15, 18, 22, 27, 33, 39, 47, 56, 68, 82, and decimal multiples and submultiples of these values. These are generally sufficient for most purposes. In analogue circuits where a precise ratio of two resistors is required (for example, to define the gain of a precision amplifier), further values are available. The E24 range includes all the E12 values plus values of 11, 13, 16, 20, 24, 30, 36, 43, 51, 62, 75, and 91. There are also E48 and E96 ranges whose values are specified to three significant figures. Generally, components from the E12 and E24 ranges will be the cheapest and should be used wherever possible.

The E ranges of component values are specified in an EIA (Electronic Industries Association) standard.

Temperature dependence

The values of electronic components often vary with temperature because the electrical properties of materials vary with temperature. For most components, the fractional change in value is roughly proportional to the temperature change and the temperature dependence of the component value can be expressed as a temperature coefficient, often in parts per million (p.p.m.) per °C.

Worked Example 5.1

A range of ¼ W resistors has a stated manufacturing tolerance of 5% and a temperature coefficient of resistance of ±200 p.p.m. $°C^{-1}$. Ignoring any other effects on resistance value, assuming the nominal value applies at 20°C, and making no assumption about the sign of the temperature coefficient, what would be the worst-case deviations from nominal value over a temperature range of 0 to 70°C?

Solution   The worst-case values at 20°C are 1.0 ±5%, or 0.95 and 1.05. The greatest deviation due to temperature will be at 70°C, where the change in resistance from the value at 20°C will be (70°C − 20°C) × (±200 p.p.m. $°C^{-1}$) or ±10,000 p.p.m or ±1%. Taking the worst-case combinations of manufacturing tolerance and temperature dependence:

Maximum deviation = 1.05 × 1.01 = 1.06 or +6%
Minimum deviation = 0.95 × 0.99 = 0.94 or −6%

Rework Worked Example 5.1, assuming that the temperature coefficient of resistance is always negative.

Stability

The electrical properties of components vary with time, whether a component is in use or in storage, due to physical and chemical changes in the materials from which the component is fabricated. Another way of looking at these changes is to say that components **age**. Component ageing can be accelerated by applied **stress**. If a component is operated continuously at its full rated voltage, for example, the component's value may change much more quickly than if the component were in storage unused. In some types of ceramic capacitors, for example, an applied voltage causes gradual changes in the crystal structure of the dielectric, resulting in a change in permittivity and hence a change in the value of the capacitor. Some other examples of stress that can accelerate ageing are: heat, which can speed up chemical changes; thermal cycling (repeatedly heating and cooling a component), which can cause joints to crack because of differential expansion; and ionizing radiation, which can disrupt the molecular and crystal structure of component materials. High-stability components have values that change comparatively little over time. Stability is expressed as a fractional change in value (usually in p.p.m. or %) over a stated time interval and under stated conditions.

Ionizing radiation, in the form of cosmic rays and subatomic particles from the Sun, is a significant problem in electronics for spacecraft.

Component ratings

Electronic components have limitations on voltage, current, power dissipation, and operating temperature range. In some cases there may also be limitations that are more complicated such as rate of change of voltage. These limitations are known collectively as **ratings**. Component manufacturers usually state two sets of ratings for their products: an **absolute maximum rating**, beyond which the component will be damaged or destroyed; and a **recommended rating**, which is the manufacturer's statement of the component's capability. To say that a capacitor has a recommended rating of 16 V does not guarantee that the capacitor will work as well at 16 V as it will at 10 V: it will almost certainly be more reliable at 10 V than at 16 V, and it may also be more stable. For these reasons, design engineers normally use a component well within its recommended rating.

There is, of course, a cost penalty in using a 16 V capacitor for a 10 V application, but the initial cost of the component may be less important than the long-term cost of unreliability.

Absolute maximum ratings may be important under fault or transient conditions. If a fault occurs in a system, other components will be undamaged if protective devices (such as an overvoltage trip circuit) operate before absolute maximum ratings have been exceeded.

Parasitic behaviour

So far in this discussion of passive component characteristics, we have looked at nonideal properties that are due to the physical limitations of materials and manufacturing processes. There is another way in which passive components can be nonideal, which is due to their electromagnetic behaviour rather than to the limitations of materials.

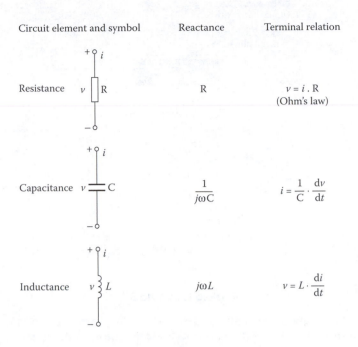

Circuit element and symbol	Reactance	Terminal relation
Resistance v R	R	$v = i \cdot R$ (Ohm's law)
Capacitance v C	$\dfrac{1}{j\omega C}$	$i = \dfrac{1}{C} \cdot \dfrac{dv}{dt}$
Inductance v L	$j\omega L$	$v = L \cdot \dfrac{di}{dt}$

Figure 5.2   The resistance, capacitance, and inductance parameters.

In lumped-parameter circuit theory, there are three simple circuit elements: resistance, capacitance, and inductance. Figure 5.2 summarizes their properties. Practical realizations of all three circuit elements exist in the form of resistors, capacitors, and inductors, but in all three cases the practical component possesses a little of the other two circuits' properties. (Note the suffixes **-ors** for a component and **-ance** for a circuit property.)

Figure 5.3a shows the construction of a metal film resistor, made by depositing a metallic film on the surface of a ceramic cylinder and then cutting a helical track into the film to obtain the desired resistance value. The helical construction of the resistive track suggests inductance, and there is also capacitance between turns of the helix. Figure 5.3b suggests a possible equivalent circuit for this type of resistor, but in reality the resistance, capacitance, and inductance of the component are physically distributed and not lumped as suggested. The series inductance, L_s, and the parallel capacitance, C_p, are known as **parasitic** properties of the resistor. In many applications, their presence can be neglected. At low frequency, perhaps in an audio circuit, the reactance of L_s is low and the reactance of C_p is high, so that the resistor behaves almost as a pure resistance. At higher frequencies, however, the impedance of the resistor is lower than at low frequencies as the reactance of C_p decreases. It is important to realize that the series inductance cannot be eliminated by making the resistor a straight bar, although it is reduced. This is because inductance is associated with the magnetic field induced by a changing current, not with a helical or spiral conductor shape. Similarly, the parasitic capacitances are associated with the electric field between two charged conductors, not with parallel flat plates.

In a lumped-parameter circuit, we model the circuit elements as if they were localized and connected by zero-impedance wires. Contrast this with a distributed-parameter circuit such as a transmission line where capacitance and inductance are distributed evenly along the line.

This point is developed further in Chapter 8.

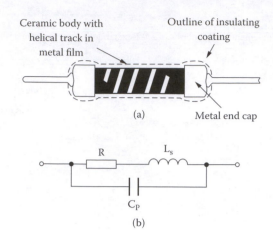

Figure 5.3   A metal film resistor: (a) construction and (b) a possible equivalent circuit.

Worked Example 5.2

Estimate the parasitic inductance of a metal-film resistor of the type shown in Figure 5.3, if the ceramic body is 3 mm in diameter and the helical track consists of 10 turns spaced over a 10 mm length.

Chapter 8 of Compton (1990) discusses calculation of self-inductance.

Solution   If we assume the ceramic has a relative permeability μ_r of 1, we can use an approximate formula for the inductance of a single-layer air-cored coil:

$$L = N^2 \mu_0 A/l$$

where N is the number of turns in the coil, μ_0 is the permeability of free space, A is the cross-sectional area of the coil, and l is its length. Hence,

$$L = \frac{10^2 \times 4\pi \times 10^{-7} \times \pi \times (1.5 \times 10^{-3})^2}{10 \times 10^{-3}} \cong 90 \text{ nH}$$

Exercise 5.3

Using the parasitic inductance value calculated in Worked Example 5.2, find out at what frequency the inductive reactance of such a resistor becomes more significant than the resistance, for resistance values of (a) 10 Ω, (b) 1 kΩ, and (c) 100 kΩ.

(*Answers*: [a] 18 MHz; [b] 1.8 GHz; [c] 180 GHz [which is so high that lumped parameter circuit models would no longer be valid, so the result is meaningless].)

Figure 5.4 shows a possible equivalent circuit for a practical capacitor. The series inductance, L_s, and resistance, R_s, are due to the wire leads of the capacitor, while the parallel resistance, R_p, is due to the leakage resistance of the dielectric. If the capacitor is of wound construction, there is an additional contribution to L_s. In reality, as with the resistor discussed earlier, the parasitic properties are distributed within

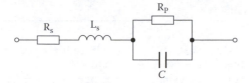

Figure 5.4   A possible equivalent circuit for a capacitor.

the capacitor to some extent. The properties of a real capacitor or inductor cannot therefore be defined analytically, and an empirical approach is taken. A capacitor or inductor is represented by a pure reactance, X, and a series resistance, R, both of which depend on frequency. The impedance of the component is then a function of frequency:

$$Z(\omega) = R(\omega) + jX(\omega) \qquad (5.1)$$

$X(\omega)$ can be recognized as the normal capacitive or inductive reactance, while $R(\omega)$ is known as the **equivalent series resistance** (ESR). Both depend on frequency. Several other quantities are derived from R and X, and all depend on frequency. The dissipation factor (DF) is the ratio of R to X and is also known as tan δ. The angle δ is called the **loss angle**, and its relationship to R and X is shown by the diagram in the margin. The reciprocal of the dissipation factor is known as Q, for quality factor. Q is defined as

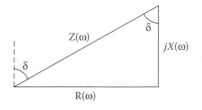

$$Q = 2\pi \times \frac{(\text{Maximum energy stored in component})}{(\text{Energy dissipated per cycle})} \qquad (5.2)$$

The concept of Q factor is also used in resonant circuits and waveguide cavities with similar meanings in terms of energy losses.

A capacitor or inductor with a high Q factor, or low dissipation factor, absorbs little power when used in an a.c. circuit. In tuned circuits, it is not possible to realize a high Q for the complete circuit unless the inductors and capacitors have high individual Q factors.

Some manufacturers may state a power factor (PF), a term that is more often used in a.c. circuit theory, rather than a dissipation factor or Q factor. If the product of the root mean square (r.m.s.) current and r.m.s. voltage flowing into a circuit is multiplied by the PF, the result is the power dissipation in the circuit in watts. In general, the current and voltage are not in phase, so that the product of their r.m.s. values (in volt–amperes, or VA) does not represent actual power dissipation.

Power factor is covered by Kip (1969), and is an important idea in electrical power engineering.

Selection of a passive component for a particular application requires a knowledge of component characteristics in general, as introduced so far in this chapter, and also an understanding of the different types of resistors, capacitors, and inductors and their fabrication and characteristics, which is what the remainder of this chapter covers.

Resistors

Resistors are used in electronic circuits for limiting current, setting bias levels, controlling gain, fixing time constants, impedance matching and loading, voltage division, and sometimes heat generation. The resistance, $R(\Omega)$, of a material of length l and cross-sectional area A is given by

Table 5.1   Characteristics of resistor types

Type	Typical tolerance (%)	Typical temperature coefficient (p.p.m. °C⁻¹)	Power rating (W)	Range of values	Operating temperature range (°C)	Features
Precision wire-wound	0.1	< 10	< 1	$10\,\Omega - 100\,\text{k}\Omega$	−55 to +145	High stability, low noise
Precision metal film	0.1	±15	⅛	$10\,\Omega - 1\,\text{M}\Omega$	−55 to +155	Low noise, good stability
Metal film	1	50–100	⅛, ¼, ½	$0.1\,\Omega - 1\,\text{M}\Omega$	−55 to +155	Low noise, good stability, lower cost than precision metal film
Carbon film	5	150–800	⅛ – 1	$10\,\Omega - 1\,\text{M}\Omega$	−55 to +155	Low cost
Power wire-wound	5	50–250	Up to 600	$0.1\,\Omega - 10\,\text{k}\Omega*$	−55 to +250	Higher-rated types may require heatsinks
Thick-film networks	2	100	⅛, ¼	$10\,\Omega - 100\,\text{k}\Omega$	−55 to +125	Multiple resistors per package
Surface mount "chips"	1	100	¼	$1\,\Omega - 10\,\text{M}\Omega$	−55 to +155	Thick or thin film on ceramic construction, low inductance

* Dependent on rating.

$$R = \frac{\rho l}{A} = \frac{l}{\sigma A} \qquad (5.3)$$

where ρ is the material resistivity (Ω m), σ is the material conductivity (Ω^{-1} m^{-1}), and l and A are expressed in metres and square metres respectively.

There are three main ways of fabricating a resistor, each applicable over a range of resistivities and resistance values (Table 5.1). First, if a material of suitable resistivity can be made, Equation 5.3 can be realized directly in terms of a slab or rod of resistive material with metal contacts at each end. If this method of construction is not feasible, the resistor can be fabricated from a longer length of material of thinner cross-section. One way of doing this is to wind a wire onto a cylinder, and the other method is to use a film, or thin layer, of the resistive material, deposited onto an insulating substrate. Slab or rod resistors for surface mounting have the lowest series inductance of any resistor types because of the absence of leads. Wound resistors tend to have high series inductance, although this can be reduced to some extent by special winding techniques in which one part of the winding cancels the inductance of the remainder.

At high frequencies, the geometric form of a resistor must be considered and special geometries such as discs may be needed.

The temperature coefficient of resistance and the stability of a resistor are determined primarily by the properties of the resistive material used to fabricate the resistor. Most pure metals have temperature coefficients of resistance of around 4000 p.p.m. °C⁻¹ and fairly low resistivities, making them unsuitable for use in resistors. Lower temperature coefficients of resistance can be achieved by fabricating resistors from proprietary alloys with temperature coefficients as low as ±5 p.p.m.

°C^{-1}. Many of these alloys are based on nickel, chromium, manganese, and copper. Well-known examples are the alloys known as nichrome (80% nickel, 20% chromium), constantan (55% copper, 45% nickel), and manganin (85% copper, 10% manganese, < 5% nickel) with temperature coefficients of < 100 p.p.m. °C^{-1}, < 20 p.p.m. °C^{-1}, and < 15 p.p.m. °C^{-1} respectively. Metal alloys are used in the fabrication of **precision** wire-wound and metal-film resistors. These types of resistor have the lowest temperature coefficients and the best stabilities of any resistor type. High-value wire-wound resistors are not feasible because of the fairly low resistivity of resistance wire. Nichrome wire of only 0.02 mm diameter has a resistance of 3.44 kΩm^{-1}, and this is about the thinnest practical wire for production resistor manufacture; over 3 m of wire is required for resistor values above 10 kΩ. The range of values that can be obtained by varying the length and pitch of a helical track in a metal film is limited, and higher-resistance values require either thinner films or films of higher sheet resistivity, which can be obtained by including nonconducting materials in the film during deposition from a vapour.

For **general purpose use** where low temperature coefficients and high stability are not essential, cheaper resistors can be fabricated using carbon as the resistive material. Carbon film resistors are fabricated by pyrolytic decomposition of a carbon-containing gas, such as methane, in a furnace, depositing a carbon film onto a ceramic or glass substrate. The resistors are then fitted with end caps and leads, and coated with a protective varnish, lacquer, or plastic.

A third type of resistive material used in resistor fabrication is known generically as **cermet**, for ceramic–metal. These materials contain finely divided metals distributed in a glassy vitrified ceramic, which can be printed or painted onto a substrate in the form of a paste and then fired to fuse the constituents into a hard solid. There are two important applications for these materials. One is in variable resistors or potentiometers where a low temperature coefficient of resistance (< 200 p.p.m. °C^{-1}) is required. The other is in **thick-film hybrid circuits** and thick-film resistor networks. Thick-film circuits consist of resistors and small-value capacitors printed and fired onto a ceramic substrate. Surface-mounted ICs and transistors can be soldered to the circuit, which can then be coated with epoxy resin and used either as a complete self-contained circuit or as a component on a conventional printed circuit board (PCB). Thick-film resistor networks are made in the same way but contain only resistors. They are especially useful where several resistors of the same value are required in the same location on a PCB. A common application is a group of eight **pull-up** resistors connected to a microprocessor bus with a common connection to a supply rail. A thick-film single-in-line (SIL) resistor pack containing eight commoned resistors has only nine terminals and occupies a smaller board area than eight discrete resistors.

Noise

All resistors generate electrical **noise**, or small random fluctuations of voltage or current. Noise is not necessarily due to material imperfections in resistors, although some types of resistor are noisier than others:

A sheet of conductor has a resistivity, measured in ohm per square ($\Omega\square^{-1}$), which is a constant for any sized square of the sheet. Metal films can be fabricated with sheet resistivities of up to about 20 k $\Omega\square^{-1}$.

Thick-film circuits are discussed in more detail by Till and Luxon (1982).

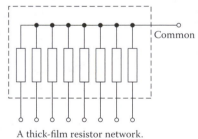

A thick-film resistor network.

Senturia and Wedlock (1993) discussed noise in more detail.

noise is a fundamental property of all conductors. Electronic systems handling weak signals have to be carefully designed to minimize the amount of noise added to the signal during processing and to achieve an acceptable **signal-to-noise ratio**. There are two types of noise generated in resistors: **Johnson noise**, which occurs in all resistors, and **excess noise**, which occurs in some types of resistor. An additional type of noise, **shot noise**, is important in devices with a potential barrier such as p–n junction diodes and junction transistors, but does not occur in resistors. Johnson noise is due to the random thermal motions of electrons in a resistor. The r.m.s. noise voltage developed in a resistor of value R at an absolute temperature T is given by

$$\sqrt{(\overline{v_n^2})} = \sqrt{(4kTRB)} \tag{5.4}$$

There is an upper frequency limit set by quantum effects. Equation 5.4 is valid for frequencies below f where $kT \ll hf$ and h is Planck's constant (6.626×10^{-34} J s). At room temperature (300 K), Equation 5.4 is valid to more than 500 GHz.

where k is Boltzmann's constant (1.38×10^{-23} J K^{-1}) and B is the bandwidth or frequency range over which the noise is measured. Johnson noise has uniform spectral density at all frequencies and is described as **white noise** by analogy with white light.

Worked Example 5.3

What is the r.m.s. noise voltage developed in a 1 MΩ resistor at room temperature (290 K) over a 20 kHz bandwidth due to Johnson noise?

Solution   From Equation 5.4,

$$\sqrt{(\overline{v_n^2})} = \sqrt{(4 \times 1.38 \times 10^{-23} \times 290 \times 10^6 \times 20 \times 10^3)}$$

$$= 18\mu\text{V r.m.s.}$$

Johnson noise depends only on resistor value and temperature and is independent of the resistor material. Wire-wound resistors exhibit only Johnson noise. Other resistor types generate **excess noise** over and above the inevitable Johnson noise. Excess noise increases with current. Carbon film resistors are the noisiest type.

Potentiometers

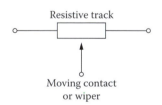

Resistive track

Moving contact or wiper

Resistors with a movable contact or tap are called potentiometers and find application in electronic circuits as operating adjustments or as manufacturing adjustments (**trimmers** or **presets**). Special potentiometers are also used as angular position transducers in servomechanisms. The resistive medium can be a wire winding or a cermet or carbon film, with the same characteristics, such as temperature coefficient, as fixed resistors of the same material. The potentiometer track can be straight with a sliding contact, or circular or helical with a rotating contact. Helical tracks are used on multiturn potentiometers for applications where fine adjustment of resistance is required. The potentiometer track may be of uniform resistance (linear) or with resistance varying with position, usually logarithmically. Power ratings are generally less than 1 W.

Table 5.2 Characteristics of capacitor types

Type	Typical tolerance* (%)	Typical temperature coefficient (p.p.m./°C)	Maximum voltage† (V)	Range of values† (F)	Operating temperature range (°C)	Features
Silvered mica	1	70	500	2 p–47 n	–40 to +85	High Q, high stability, high cost
Ceramic	10	100–750	500	1 p–1 μ	–55 to +125	Surface mount or leaded
Polystyrene	2	< 200	500	10 p–10 n	–40 to +85	Low leakage (> 10^{12} Ω)
Polycarbonate	5	50	400	10 n–10 μ	–55 to +125	
Polyester	10–20	200	100	1 n–1 μ	–55 to +125	Surface mount or leaded
Polypropylene	5	200	> 1500	1 n–1 μ	–55 to +100	Low leakage (> 10^{11} Ω)
Solid tantalum	10–20	—	35	10 n–300 μ	–55 to +85 or +125	Polarized, leakage depends on value
Aluminium (Al) electrolytic	20	—	400	1μ – > 100,000 μ and higher	–40 to +85	Short life, wide temperature variation (e.g., 4 years/3600 hours)

* An absolute tolerance often applies for values below 10 pF.

† High values and high maximum voltage are not available simultaneously.

Capacitors

Capacitors of various types (Table 5.2) are used in electronic circuits for storing charge, as elements of frequency-selective circuits and filters, for coupling a.c. signals from one circuit to another, and for shunting unwanted signals to ground (decoupling). Most capacitors used in electronic engineering are of the parallel-plate type whose capacitance, C, in farads (F), is given by

$$C = \varepsilon_0\varepsilon_r A/d \qquad (5.5)$$

where A is the area of the plates (m²), d is the plate separation (m), ε_0 is the permittivity of free space, and ε_r is the relative permittivity of the dielectric medium between the plates (dimensionless). An alternative way of quantifying the multiplying effect of the dielectric is to state the dielectric constant, K, of the material where $K = \varepsilon_0\varepsilon_r$. The properties of dielectrics have a significant influence on the properties of capacitors and will be briefly described.

Dielectrics

Equation 5.5 shows that the presence of a dielectric between the plates of a capacitor increases the capacitance by a factor of ε_r over the same capacitor with air between the plates. The increased capacitance due to the dielectric is caused by **polarization** of the dielectric in the presence of an electric field. Polarization is a slight shift of the negative charge in the dielectric relative to the positive charge, which causes a net charge to appear on opposite faces of the dielectric. In a capacitor, the

A farad (F) is 1 coulomb per volt. The capacitance of a capacitor, C, charged to V volts by a charge Q coulombs, is given by $C = Q/V$.

Since $\varepsilon_r \approx 1.0006$ for air, we can regard air as almost equivalent to a vacuum in terms of permittivity.

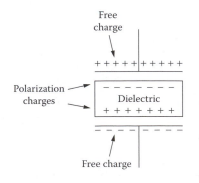

87

polarization charges neutralize some of the free charge on the capacitor plates and reduce the potential difference between the plates. Since capacitance is defined as Q/V, a reduction in the potential difference, V, implies an increase in capacitance. Put another way, because the polarization charges neutralize some of the free charge stored on the plates, the capacitor is able to accept more stored charge for a given potential difference between the plates.

There are several different mechanisms at the atomic or molecular level that can contribute to the polarization and therefore the relative permittivity of a dielectric. Each mechanism has its own characteristic time constant, and the relative permittivity of a dielectric is therefore frequency dependent. The slowest polarization mechanism that contributes to the relative permittivity of a dielectric at low frequencies only is the migration of free electrons and impurity ions in the presence of an electric field. There are two polarization mechanisms at molecular level that can operate in materials formed from or containing polar molecules. One is orientation of the polar molecules in the direction of the applied field; the other is stretching or distortion of the polar molecules in the direction of the field. The fastest polarization mechanism, which operates at optical frequencies, is displacement of the electron clouds in atoms of the material relative to their nuclei.

When a dielectric is polarized by an alternating electric field, energy losses occur in the dielectric. These energy losses generate heat and are responsible for the resistive component $R(\omega)$ of the complex impedance of a capacitor in Equation 5.1. The loss angle, δ, of a capacitor can be seen to be a property of the dielectric.

Polymer films used in capacitor manufacture have relative permittivities between 2 and 5. Metal oxides can have values as high as 80 (titanium dioxide), and some ceramics, notably barium titanate, can have relative permittivities of up to 12,000. The latter are known as **ferroelectric** or **high-K** ceramics, and although they can be used to fabricate compact capacitors, the capacitance values are not very stable.

The maximum voltage that a capacitor can withstand without damage is determined by the dielectric strength of the dielectric (in V m^{-1}) and by the thickness of the dielectric. For a given dielectric material, capacitors of higher working voltage can be made by using a thicker dielectric, but as Equation 5.5 shows, increasing the dielectric thickness, d, decreases the capacitance. The physical size of capacitors therefore increases with capacitance value and with working voltage.

Dielectrics are not perfect insulators, and some current will flow between the plates of a capacitor with a steady voltage applied. Capacitors, therefore, have a d.c. **leakage resistance**, which for the best capacitor types is as high as 10^{12} Ω for a capacitance of around 10 nF. Apart from a steady leakage current, capacitors also exhibit **dielectric absorption** in which charge is absorbed into the dielectric. A capacitor that has been short-circuited and supposedly fully discharged can recover a small voltage as the absorbed charge emerges from the dielectric. This can be a problem in precision analogue-to-digital converters, where the result of an A-to-D conversion can be slightly influenced by the result of the previous conversion because the capacitor has some "memory" of its previous voltage. It can also be a safety problem with

In a polar molecule, there is an asymmetric distribution of positive and negative charge, and the molecule has a dipole moment.

Polarization is discussed by Kip (1969), Kraus (1992), Carter (1992), and Compton (1990). A more detailed discussion of polarization and dielectrics is given by Anderson et al. (1990).

Ferroelectric materials are discussed by Anderson et al. (1990).

Leakage resistance less than this can be created along the outside of a capacitor by contamination with flux or even oils from the human skin.

large-value, high-voltage capacitors. This point is considered further in Chapter 11.

Capacitors can be divided broadly into two types: those that have a solid dielectric of ceramic or polymer, and those that have a thin metal-oxide film as a dielectric. For reasons that will be seen later, the latter are known as **electrolytic capacitors**; and, for want of any better name, we can refer to the remaining types as **nonelectrolytic**.

Nonelectrolytic capacitors

Nonelectrolytic capacitors used in electronic engineering fall into three main groups: mica, ceramic, and polymer film. Mica capacitors are fabricated from thin sheets of naturally occurring mica, a laminar aluminium silicate that can be split into sheets as thin as 25 μm. The capacitor plates are formed from a silver coating on each side of the sheet. Silvered mica capacitors are stable and have a high Q factor. They are used for this reason in resonant circuits. Because mica cannot be rolled up, there is a practical limit of about 100 nF on capacitor value, and, except in small values, multiple sheets of mica have to be stacked together to obtain the required capacitance.

Exercise 5.4

Estimate the maximum capacitance value achievable by stacking sheets of mica 25 μm thick measuring 30 mm square if a maximum of 10 sheets can be stacked. ε_r for mica is about 6 and ε_0 is about 9 pFm^{-1}.

(*Answer*: 20 nF.)

Ceramic capacitors are made either as flat discs or as rectangular multilayer monolithic blocks depending on the value required. The ceramics used are mainly based on titanium dioxide (TiO_2) or barium titanate ($BaTiO_3$). Other compounds are added to the ceramic to give specific electrical properties. Dielectrics based on barium titanate have very high relative permittivities and are known as high-K dielectrics. High-K ceramic capacitors have poor stability and are thus best suited to applications such as decoupling, where a stable capacitance value is not essential but where the compactness of a high-K capacitor is useful. Ceramic capacitor electrodes are formed from silver, platinum, or palladium, which is normally applied as a paste and then fired to fuse with the ceramic. If leads are required, they are then soldered to the electrodes, and the capacitor body is coated to protect the solder joints. Ceramic "chip" capacitors without leads for surface mounting need no protective coating and give excellent high-frequency performance because of the lack of leads and the consequent low series inductance.

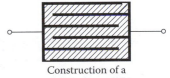

Construction of a multilayer monolithic ceramic capacitor.

Polymer film capacitors are made either by coating a plastic film with metal by vapour deposition or by forming a sandwich of metal foil and plastic film. The film or foil is then rolled into a cylinder with either axial or radial leads. Polymer capacitors tend to be bulky for their capacitance because of the low relative permittivity of polymers. A variety of polymers are used in capacitor construction, of which only the most common are included in Table 5.2.

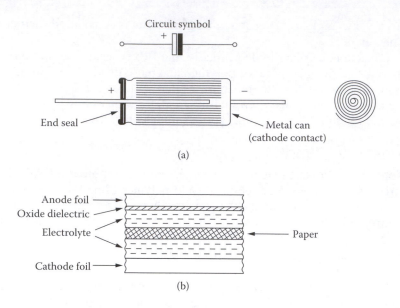

Figure 5.5   Construction of an electrolytic capacitor: (a) general arrangement and (b) detail of foils, dielectric and electrolyte.

Electrolytic capacitors

As we have seen from Equation 5.5, large-value capacitors must have a dielectric of high relative permittivity and minimum thickness together with a large plate area. Electrolytic capacitors combine all of these factors and achieve the lowest volume, for their capacitance and voltage rating, of any capacitor type. They are used in circuit applications requiring large values of capacitance, such as power-supply circuits, where they find use as reservoir capacitors, low-frequency filters, multistage amplifiers, and timing circuits with long time constants.

The construction of a typical electrolytic capacitor is shown in Figure 5.5. The capacitor plates are metal foil, and the dielectric is a thin film of metal oxide formed on one of the plates, called the anode. The dielectric is connected electrically to the other plate (the cathode) by an electrolyte or ionic conductor. The cathode and dielectric are also separated by porous paper sheets to prevent direct contact between the cathode and dielectric. The paper does not prevent current flow because of its porosity. The type of electrolytic capacitor just described is **polarized** and must be connected in a circuit with the anode at a more positive potential than the cathode. This type of electrolytic capacitor is the most common and has the greatest capacitance for its volume of all electrolytic types. If a polarized electrolytic capacitor is subjected to reverse polarity, either the dielectric will break down or an oxide film will form on the cathode, reducing the capacitance. Nonpolarized electrolytic capacitors are manufactured with oxide films on both electrodes for a.c. circuit applications. Their capacitance is lower than for a polarized type of the same volume, for the reason just mentioned.

The two most common electrolytic capacitor types are made from either aluminium or tantalum. Aluminium electrolytics consist of two aluminium foil electrodes, one of which has an aluminium oxide film

The charge carriers in an electrolyte are mobile **ions** or ionized atoms that have either gained or lost one or more electrons. An electrolyte need not be liquid.

An oxide film on both electrodes is equivalent to two capacitors in series. Electrolytic capacitors may explode if subjected to reverse polarity.

formed on its surface, separated by a wet or paste electrolyte. They are characterized by a wide tolerance on initial value, considerable capacitance variation with temperature, and a short life at high temperatures. Their capacitance decreases with age. Tantalum electrolytic capacitors can be made with foil electrodes and wet electrolyte in the same way as aluminium electrolytics. They can also be made with a solid dry electrolyte, achieving long life and high reliability at reasonable cost. The dielectric film is tantalum pentoxide, formed in a thin layer and with large surface area by sintering tantalum powder into a solid slug or bead anode that is coated with carbon and then metal-plated to form the cathode contact.

Exercise 5.5

Electrolytic capacitors are often used for power rail decoupling on PCBs to shunt any interference present on the rail. Often, circuit designers add a small ceramic capacitor as shown in parallel with the electrolytic. The value of the ceramic capacitor might be only 100 nF, while the electrolytic could be of 100 μF or more. Capacitances in parallel add, so the total capacitance of the combination shown would be 100.1 μF. What is the purpose of the ceramic capacitor despite its negligible effect on the total capacitance?

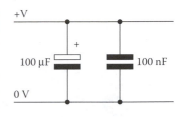

Inductors

In most low-frequency electronic circuits, inductors are usually avoided altogether because of the expense and inconvenience of manufacture and because of the better control over circuit characteristics obtainable by using capacitors as the reactive elements. At higher frequencies, inductors are used as elements of tuned circuits and filters. At microwave frequencies, inductance can be obtained at minimal cost by inductive conductor patterns on circuit boards and in integrated circuits (ICs). Power inductors are used in switched-mode power supplies as described in Chapter 4. Most inductors have to be specially designed and wound for each application, although there are a few off-the-shelf inductors available with values from 1 μH to 1 mH, including surface mount versions in the smaller values, such as 10 nH.

Inductors are still referred to as **chokes** in some applications.

Although transformers are not discussed here, much of what follows about inductor manufacture is also relevant to the fabrication of small transformers for signal coupling, isolation, and impedance matching.

There is no simple equation for the inductance of an inductor analogous to Equation 5.5 for parallel plate capacitors because inductor geometries vary considerably depending on the method of manufacture. For the **special case** of an N-turn inductor wound on a toroid of a suitable material, the inductance is

$$L = N^2 \mu_0 \mu_r \frac{A}{l} \tag{5.6}$$

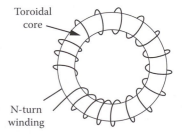

Toroidal core

N-turn winding

where μ_0 is the permeability of free space ($4\pi \times 10^{-7}$ H m^{-1}), μ_r is the relative permeability of the toroid material (dimensionless), A is the cross-sectional area of the toroid (m^2), and l is the mean circumference of the toroid (m). The toroid forms a magnetic circuit that contains the magnetic flux generated by a current in the winding. In other geometries the

Compton (1990) and Carter (1992) discussed calculation of inductance in more general cases.

magnetic flux is not totally contained within the magnetic circuit and Equation 5.6 does not apply. Nevertheless it does give some insight into inductors in general. Inductance is proportional to the square of the number of turns, N, in the winding, whereas the resistance, volume, and weight of the winding are proportional to N. The cross-sectional area, A, of the magnetic circuit should be large to obtain a large value of L, but the length of the magnetic circuit, l, should be kept small. Small, fat toroids are preferred therefore (but with only a small hole through the middle, there will not be much space for windings). As in the case of capacitors where a dielectric of high relative permittivity is desirable to obtain large values of capacitance, in inductors the core material should have high relative permeability. Ferromagnetic metals such as iron, nickel, cobalt, and their alloys have values of μ_r as high as 200,000. Inductors with metal cores tend to have high losses because of circulating **eddy currents** in the core. Low-loss inductors can be fabricated using **ferrites** as the core material, which have high μ_r but are electrically insulating. Ferrite components for inductors are widely available commercially. Manufacturers give detailed empirical data on inductor design using their components. The smallest ferrite components are **ferrite beads** that are designed to thread onto a wire to form a single-turn inductor. They are used to control parasitic signal pickup and to filter unwanted high-frequency interference.

A ferrite bead.

Relays

Electromagnetic relays are still used in some application areas of electronics. They consist of a coil with an iron core and a movable iron pole-piece that operates one or more sets of switch contacts. The coil is isolated from the switch contacts so that a low-power circuit driving the coil can switch an independent circuit, possibly carrying much larger currents or voltages. In many electronic applications, **reed relays** are often used. They consist of a **reed switch** in a glass envelope, fitted axially into a solenoid as shown in Figure 5.6b. Reed relays are capable of fast switching, and can often be driven directly by logic circuits. A relay coil is, of course, an inductor, and the transistor circuit shown in Figure 5.6c includes a **freewheel diode** to provide a path for the current generated by the coil as the magnetic field collapses after the transistor has turned off.

Summary

In this chapter, the differences between real components and ideal circuit elements have been examined, concentrating on passive components. All passive components are subject to manufacturing variations that are quantified by an engineering tolerance on a component's value. Component values vary with temperature and with age. Temperature variation is normally expressed using a temperature coefficient, and the effect of age is expressed as a component's stability.

All components have parasitic properties due to their electromagnetic behaviour and their materials. The parasitic properties include unwanted reactance and resistance, energy losses in capacitors, and inductors and noise in resistors.

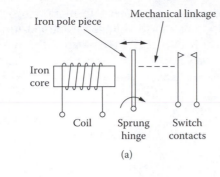

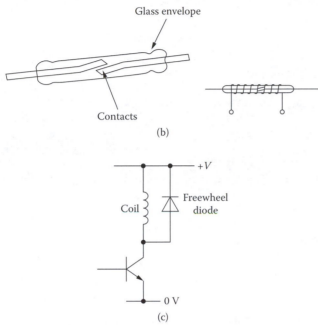

Figure 5.6   Electromagnetic relays: (a) conventional type, (b) reed type, and (c) transistor drive circuit.

The main types of resistor and capacitor used in electronics have been introduced, and inductors have been discussed briefly.

Problems

5.1   The voltage gain of a particular precision amplifier is set by the ratio of two resistors and must be accurate to within ±2.5% of the design value over the full operating temperature range of the amplifier (0–40°C) and over the full expected life of 5 years. The designer chooses metal film resistors with a temperature coefficient of ±50 p.p.m. °C^{-1} and with a stability of ±1% of initial value over 5 years. What initial tolerance on the resistor values at 20°C is needed?

5.2   If a 1 kΩ resistor is to be used at 50 MHz, what is the maximum acceptable parasitic shunt capacitance of the resistor if the capacitive reactance is to be no fewer than 10 times the resistor value?

5.3 Low-leakage capacitors can have insulation resistances of 10^{12} Ω. What is the r.m.s. noise voltage developed in a resistance of this value at 300 K over a 10 kHz bandwidth? Why are practical capacitors not noisy?

Instruments and measurement

6

Objectives

☐ To review the fundamentals of electrical measurements important in electronic engineering.

☐ To explain the principles of operation of common voltage, current and frequency measuring instruments.

☐ To explain the principle of operation of the oscilloscope.

☐ To introduce the limitations of measuring instruments and the problem of parasitic loading of the circuit being measured.

Electronic engineering depends on the processing and transmission of signals and information in electrical (and optical) form. The electronics engineer is almost totally dependent upon instruments to observe and measure what is happening electrically inside electronic circuits, and a sound understanding of instruments and measurement is a necessary part of the engineer's set of skills. Instruments are used for many reasons and in different ways during the life of an electronic design. Initially, during design and development, **prototypes** may be constructed of the whole or part of an electronic product or circuit, and measurements and observations will be made to determine whether the circuits perform the intended function and, if so, whether the performance is adequate. At this stage the instruments used are likely to be manually operated **bench** instruments. Later, when a design is put into production, instruments will be used again to verify that each unit produced operates correctly and that its performance is within acceptable limits. This will be done using **automated** instruments or **test equipment** in most cases, the degree of automation (and cost) increasing with the volume of product to be tested. A third area where instruments will be used is **service and repair**, where units have developed faults or require recalibration. This requires observation of faulty behaviour or measurement and adjustment respectively. **Observation** is typically carried out with an **oscilloscope** so that **waveforms** within a circuit can be "seen." A great deal of useful information can be obtained about a faulty circuit by looking at waveforms. This is not strictly a "measurement," but is of fundamental importance in electronic engineering, and the operation and use of the oscilloscope are discussed at length later in this chapter.

Electronic instruments are, of course, widely used outside the field of electronic engineering to measure physical quantities such as temperature, pressure, displacement, velocity, and acceleration. These measurements depend on **transducers** to produce an electrical signal proportional to the measured physical quantity. The use of instruments in this way falls outside the scope of this book, but details can be found elsewhere. The subject of **precision measurement** is also outside the scope of this book — we are concerned here with routine measurements

See, for example, Bannister and Whitehead (1991).

such as a design engineer or test engineer might make, not with the type of measurement made in a standards laboratory to calibrate a measuring instrument against a voltage or current standard.

Much of the content of this chapter relates to analogue measurements, but of course, digital circuits and systems are dominant in many applications of electronics. Although specialized instruments, such as logic analyzers, exist for work on digital circuits, analogue measurements are still important to check signal quality. A second area where specialized equipment is used is radio-frequency or r.f. circuits. This is not covered in this book as the topic is too specialized.

Quantities to be measured

We now turn to the major electrical quantities that an electronic engineer might need to measure and review the definitions of the quantities and typical examples where each would occur. Typical instruments for measuring these quantities will then be described.

Voltage

Voltage is probably the most frequently measured quantity in electronic engineering, and the need arises to measure both steady (or d.c.) voltages and time-varying voltage, of which sinusoidally varying (or a.c.) voltages are a very common case. Voltage is a measurement of electric **potential** — it indicates the potential energy of electric charge and is expressed in units of energy per unit of electric charge. Voltage is therefore expressed in joules per coulomb ($J\ C^{-1}$), which, for convenience, is given the name volt (V). Voltage is a relative quantity — it must be measured relative to some **reference potential**. Usually the reference potential is denoted by 0 V and is quite often "mains" earth potential, that is, the potential of a conductor that is electrically connected by a low impedance to the body of the earth through the fixed wiring of a building or other location. Within an electronic circuit the 0 V rail may or may not be electrically connected to mains earth. The choice is not arbitrary and, among other things, safety must be considered, as discussed further in Chapter 11. Typically voltages from about 1 mV to 1 kV are routinely measurable with common instruments. Above a few kilovolts, care has to be taken to avoid flashover and special instruments are needed. Below 1 mV, electrical noise becomes a problem and measurements are not so straightforward.

Instruments may be calibrated to read either r.m.s. or peak values.

Time-varying voltages can be described by their peak-to-peak value or by their root mean square (r.m.s.) value. The r.m.s. value is most useful because it indicates the equivalent steady voltage that would dissipate the same power in a given resistance. The measurement of voltage requires that the measuring instrument be placed in parallel with the circuit to be measured.

Current

The need to measure current directly arises less often than the need to measure voltage. Quite often it is sufficient to estimate a current in a circuit by making a measurement of voltage across a resistor. Apart from the uncertainty in the voltage measurement, the value of the resistor may not be known accurately, introducing further uncertainty. Current is a measure of the quantity of electric charge passing a point in a circuit per unit of time. Thus, a current is expressed in coulombs per second ($C s^{-1}$), which is given the name ampere (A). The measurement of current requires that the measuring instrument be placed in series with the circuit — this is not always easily done since a conductor will have to be disconnected — hence the common use of a voltage measurement across a resistor already in the circuit.

Frequency, phase shift, and time delay

Frequency applies to time-varying signals (often in the form of a voltage) and indicates the number of cycles of the signal per second. The International System unit of frequency is the hertz (Hz) with dimensions of $second^{-1}$. Frequency is the reciprocal of **period**, and it is often measured by measuring the period of a signal, using, for example, an oscilloscope and calculating the reciprocal. If the period is denoted by T (in seconds) as shown in the margin, the corresponding frequency f (in Hz) is given by $1/T$.

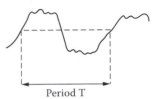
Period T

Frequency may also be used in digital circuits to denote the rate at which the circuit is **clocked** or the rate at which data are transmitted serially over a communications link.

Phase shift can be regarded as a measure of time delay, expressed in terms of the relative phase angle between two waveforms (usually sinusoidal) as defined in Figure 6.1. This arises in the characterization of **linear** circuits (such as amplifiers) where the output signal is often of different amplitude to the input signal. **Time delays** are often measured in digital circuits, for example where a signal propagates through a logic gate or chain of gates, and it is desired to measure or check the propagation delay.

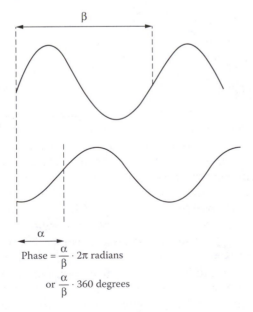

$$\text{Phase} = \frac{\alpha}{\beta} \cdot 2\pi \text{ radians}$$

$$\text{or } \frac{\alpha}{\beta} \cdot 360 \text{ degrees}$$

Figure 6.1   Definition of a phase angle.

97

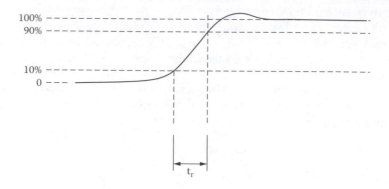

Figure 6.2   Definition of rise time.

Rise and fall time

An important quantity in digital or other circuits where a signal changes sharply from one steady level to another is the **rise time** of the signal. This is defined in Figure 6.2 and is usually measured with an oscilloscope, as described later. The corresponding quantity for a signal changing from a higher to a lower level is defined similarly and is called **fall time**.

Voltage and current measurement

There are two main types of instruments used to measure voltage and current: electronic meters (usually with a digital display) and electromechanical meters.

Electromechanical meters are inherently current-measuring devices, while electronic meters tend to be voltage measuring since they are based on analogue to digital conversion. However, both types of meter can be made to measure current or voltage by the addition of suitably arranged resistors as described below.

Electronic multimeters

Electronic meters are based on analogue-to-digital conversion, comparing a voltage to be measured against an internal reference voltage. These types of meter are usually multiple purpose and capable of measuring resistance as well as a.c. and d.c. voltage and current. They are known as digital multimeters or DMMs. Multiple voltage ranges are provided by switchable attenuators, switched automatically on more expensive **auto-ranging** instruments, or manually on lower-cost models. Current is measured by passing the current through a switch-selectable precision resistor and measuring the voltage developed across the resistor. A.c. measurements are usually accommodated by a **precision rectifier** circuit. All quantities to be measured by a digital multimeter are therefore converted to a proportional d.c. voltage, which is then converted to a digital reading by an analogue-to-digital converter (ADC). The result is then indicated, usually on a liquid crystal display. The ADC

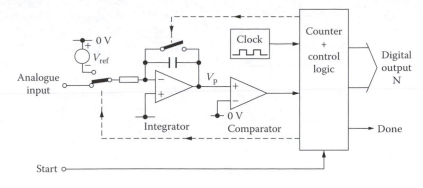

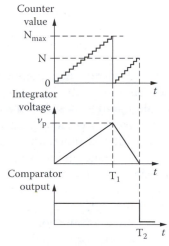

Figure 6.3   A dual-slope integrator ADC.

in most DMMs is a dual-slope integrator, which inherently measures
the **mean** value of its input voltage. This type of ADC is illustrated in
Figure 6.3 and operates in two stages. Initially the integrator is held at
zero by the switch across the integrator capacitor, and the counter is
held at zero. When the ADC is started, the analogue input is connected
to the integrator, which ramps up as the counter increments towards its
maximum value N_{max}. The first stage of the conversion ends at time T_1
as the counter overflows. During the second stage of the conversion, the
integrator is connected to the negative reference voltage V_{ref} and ramps
down towards zero. At time T_2, the comparator switches and the counter
is stopped. The final value in the counter, N, is proportional to the time
$(T_2 - T_1)$ taken to ramp the integrator down to zero, which is proportional to the **mean value** of the analogue input during the first stage of
conversion. Correct measurement of r.m.s. values for **sinewave inputs**
may be achieved by designing the precision rectifier circuit to have a
gain of $\pi / \sqrt{2}$ or about 2.2. (This figure arises because the r.m.s. value
of a sinewave is $1/\sqrt{2}$ times the peak-to-peak value, and the mean
value of a rectified sinewave is $2/\pi$ times the peak value.) Correct readings will not be obtained for nonsinusoidal a.c. voltages and currents
because of the different ratio of r.m.s. to mean value for nonsinusoidal
waveforms.

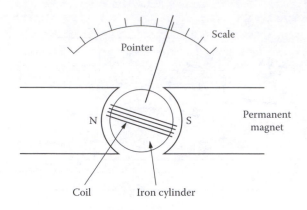

Figure 6.4   General arrangement of a moving-coil ammeter.

Electromechanical meters

Electromechanical meters depend on the deflection of a current-carry-ing conductor, normally a coil, in a magnetic field to move a pointer or needle against a calibrated scale. Figure 6.4 shows the typical arrange-ment of a **moving coil** d.c. ammeter (or current meter). A coil of wire is wound on an iron cylinder, which is mounted on axial pivots and restrained by springs so that it can rotate and move a pointer against a scale. The coil is placed between the poles of a permanent magnet with curved pole faces so that the magnetic field is radial to the cylindri-cal core. When a current passes through the coil, a torque is produced that deflects the coil through an angle proportional to the current. The restraining springs provide a balancing torque and return the coil to the rest position when the current is removed. Meters of this type are capa-ble of **sensitivities** down to 100 µA full-scale deflection (f.s.d.). Larger currents are measured by placing a **shunt resistor** or just "shunt" in parallel with the meter so that a fraction of the current being measured passes through the meter coil and the rest passes through the shunt.

Worked Example 6.1

A 1 mA f.s.d. moving coil meter with a coil resistance of 75 Ω is to be shunted with a suitable resistor to give an f.s.d. of 500 mA. Calculate the required resistor value and power rating.

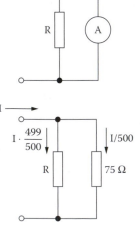

Solution   The marginal sketch shows the meter and shunt and the equivalent circuit. 1/500th of the current to be measured, I, must pass through the 75 Ω meter coil. The remaining 499/500ths must pass through the shunt of resistance, R. Applying Ohm's law and equating the voltages across the shunt and the meter coil:

$$\frac{499}{500}IR - \frac{I}{500}75$$

from which $R = 75/499$ or 0.15 Ω.

(Alternatively, the current must divide in the ratio 1:499, therefore the resistances must be in the ratio 499:1.)

To measure voltage, a small current must be drawn from the circuit being measured by placing a resistance in series with the meter such that the current drawn at full-scale voltage equals the full-scale sensitivity of the moving coil meter.

Exercise 6.1

Calculate the series resistor value required to convert a 1 mA f.s.d. ammeter with a coil resistance of 75 Ω into a 1 V f.s.d. voltmeter.

(*Answer*: 925 Ω.)

Ideally, a voltmeter would draw zero current from the circuit being measured. As shown above, a voltmeter based on a moving-coil ammeter must draw some current in order to work, and it is common to specify a **figure of merit** for this type of voltmeter. This can be done by stating the resistance of the meter divided by the full-scale deflection in volts, giving a figure in ohms per volt. A good moving coil voltmeter will typically have a figure of merit of 20,000 Ω/V. Voltmeters with a low figure of merit may draw significant current from a circuit being measured, leading to error in the measurement.

Ammeter and voltmeter functions are often combined in a multipurpose instrument known as a **multimeter**, usually with switch-selectable ranges for both a.c. and d.c. current and voltage and ranges for measuring resistance (using an internal battery to supply a current through the resistance to be measured). A.C. measurements are obtained using rectifiers (usually full-wave bridges) to obtain a unidirectional current through the meter coil. The meter responds to the mean current, but the scale may be calibrated in r.m.s. units. This will only be correct for a **sinusoidal** voltage or current within the frequency range of the multimeter. Correct measurement of nonsinusoidal currents requires a true r.m.s meter such as a **moving iron** meter. Details may be found elsewhere in, for example, reference handbooks on electrical measurement and instruments. Figure 6.5 shows the well-known "Avometer" multimeter, which has been produced more or less in its present form since 1936, and was still in production in 2006, at the time of writing this third edition. It measures a.c. or d.c. voltage and current, and resistance.

Frequency and time measurement

Rough measurements of frequency and time can be made with an oscilloscope, as discussed later in this chapter. There are relatively inexpensive instruments available, however, designed specifically for this purpose and capable of much better accuracy and precision than the oscilloscope.

Often these instruments can also count pulses or cycles of an input waveform (which can be useful in testing some digital circuits). These instruments are known as digital frequency meters, or counter/timers. A block diagram of a simple frequency meter, capable of measuring frequency and period, is shown in Figure 6.6. Although the instrument is digital, the input signal need not be. To allow for a range of input signal amplitudes, the input may be amplified or attenuated. A Schmitt trigger circuit, with a variable threshold level, then produces a digital

Figure 6.5   The Avometer Model 8 Mark 7 multimeter.
(Courtesy of Megger Limited.)

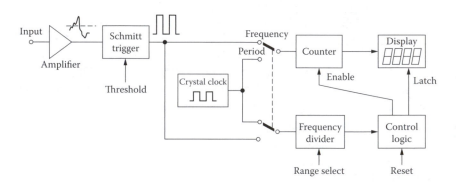

Figure 6.6   Simplified block diagram of a digital frequency meter.

The Schmitt trigger circuit is
described by Senturia and
Wedlock (1993), and in many
other texts on circuits.

waveform with a frequency equal to that of the input signal, **provided**
that the threshold level is correctly set. Figure 6.7 shows how an incorrect threshold level can cause an erroneous frequency measurement.
Some instruments have automatic control of the triggering level but are
still prone to give false frequency readings if the input signal amplitude is too small or too large. A further problem can be excessive noise
superimposed on the input waveform, which may be large enough to
trigger the Schmitt trigger and therefore give a falsely high frequency

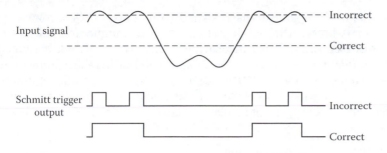

Figure 6.7 Correct and incorrect settings of the threshold level in a digital frequency meter.

reading. It is perhaps wise to examine the signal being measured with an oscilloscope, unless the signal is known to be relatively noise-free.

The digital output from the Schmitt trigger circuit is connected to a counter (for frequency measurements) or a frequency divider (for period measurement). The counter is connected to a digital display so that the displayed value from one measurement can be held while the next measurement is in progress. This is important because high precision measurements can take several minutes. In modern instruments some of the functions may be implemented in software running on a microprocessor or microcontroller, including the driving of the display.

To measure the frequency of the input signal, the counter is connected to the output of the Schmitt trigger and a precise, stable crystal clock of a suitable frequency such as 1 MHz is divided down and used to enable and disable the counter. To see how this operates, consider an input frequency of 15.625 kHz and assume that the crystal oscillator has a frequency of 1 MHz. If the crystal oscillator output is divided by 10^6 and the counter is cleared and then enabled for 1 second, the counter will register 15,625 and will thus display 15,625 Hz, or 15.625 kHz. Usually there will be a display of Hz or kHz with appropriate positioning of the decimal point on the display, operated by the range selection switches. More expensive instruments with auto-ranging automatically select the correct units and position for the decimal point.

To measure period the counter is connected to the output of the crystal clock and thus counts in units of 1 μs assuming, as above, a 1 MHz oscillator. The frequency divider is connected to the output of the Schmitt trigger. If we again assume an input frequency of 15.625 kHz, which has a period of 64 μs, and that the frequency divider divides by 10, the counter will be enabled for 10 cycles of the input waveform and will count to 640, which, with suitable positioning of the decimal point, will give a reading of 64.0 μs. Clearly, the value displayed is an **average** over 10 cycles of the input signal, and a more accurate result will be obtained by averaging over a greater number of cycles.

The accuracy of a digital frequency meter is clearly dependent on the crystal oscillator. This may require some time to warm up after the instrument is switched on (typically 15 minutes or so), and it will also be affected by temperature and ageing, which were discussed in the previous chapter. Greater accuracy may be achieved using a **frequency standard** (a precise oscillator of high stability) or a **frequency receiver**, picking up a broadcast radio signal from an atomic clock, and many

A digital frequency divider consists of a counter clocked by the signal to be divided. Division by powers of 2 is easily achieved with a binary counter, but other ratios are possible.

In the UK the carrier frequency of the BBC's 198 kHz longwave transmitters is maintained to an accuracy of ±2 parts in 10^{11} with a daily variation less than 1 part in 10^{11}, but the carrier must be averaged over more than 1 second because it is phase modulated with low frequency data. In the United States, the National Institute of Standards and Technology (NIST) broadcasts frequency standard signals at 60 kHz and at 2.5, 5, 10, and 15 MHz. The global positioning system (GPS) also offers accurate frequency reference signals.

instruments have an external socket for this purpose. A second source of error is inherent in the digital counting technique and is known as **±1 count error**. Whether measuring frequency or period, the counter may or may not count a final pulse at the end of the measurement, and repeated measurements may differ by 1 in the least significant position. (**Note**: this may cause 1 kHz to be measured as 999 Hz and does not just affect the least significant displayed digit.)

Waveforms — The oscilloscope

The oscilloscope is probably the most widely used and indispensable instrument available to an electronics engineer. It allows measurements of voltage, frequency, phase, and time to accuracies of about 2%, but more importantly, it allows waveforms to be visualized and the operation of a circuit to be observed. Conventional analogue oscilloscopes can display repetitive waveforms or signals only. Digital oscilloscopes are used to capture one-off or transient waveforms for display, and are usually also capable of displaying periodic waveforms in the same way as a conventional oscilloscope. (This is known as **real-time** display.) Most oscilloscopes sold today are digital, but some analogue instruments are still available. Oscilloscopes are characterized mainly by their **bandwidth**. A typical low-cost bench oscilloscope has a bandwidth from zero to 20 or 50 MHz, while more expensive instruments are available with bandwidths of up to 400 MHz. The upper frequency limit is usually that at which the displayed waveform is smaller in amplitude than the true value by 3 dB. Oscilloscopes of lower bandwidth will not display a faithful representation of waveforms with significant frequency content outside the bandwidth. This is particularly important when displaying digital signals with fast rise and fall times.

A voltage ratio of –3 dB corresponds to a reduction in amplitude by a factor of $\sqrt{2}$.

Worked Example 6.2

Sketch the waveform that would be displayed on a 50 MHz bandwidth oscilloscope if the input waveform was a 10 MHz square wave of 50:50 mark:space ratio.

Fourier Series are discussed by Meade and Dillon (1991).

Solution   The Fourier series for a square wave of amplitude $\pm V_p$ is

$$\frac{4}{\pi}V_p\sum_{n\ \text{odd}}\frac{1}{n}\sin n\omega_0 t$$

where ω_0 is the fundamental frequency in rads^{-1}. Assuming a unit amplitude 10 MHz square wave, this is composed of a 10 MHz fundamental with amplitude $4/\pi$ and harmonics at 30 MHz with amplitude $4/3\pi$, at 50 MHz with amplitude $4/5\pi$, and so on. We may assume that the harmonics above 50 MHz will not affect the display because they are well outside the oscilloscope bandwidth, thus leaving us with three sinusoidal components at 10 MHz, 30 MHz, and 50 MHz. Of these, the 50 MHz component is reduced in amplitude by 3 dB (to about 70% of its "true" amplitude) when displayed because it coincides with the upper bandwidth limit of the oscilloscope.

Thus the amplitudes of the harmonics **as displayed** are:

10 MHz $4/\pi$ $\simeq 1.27$
30 MHz $4/3\pi$ $\simeq 0.42$
50 MHz $4/5\pi \times 0.7 \simeq 0.18$

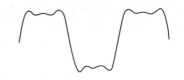

and adding together three sinewaves with these amplitudes and frequencies gives the waveform shown in the margin.

Figure 6.8 shows a simplified block diagram of a typical analogue two-channel oscilloscope. The main component is the cathode ray tube or CRT. This is an evacuated glass tube with a flat front face coated on the inside with a **phosphor** that emits light when struck by an electron beam. An electron gun at the back of the tube generates a collimated beam of electrons by thermionic emission from a hot wire filament. The electrons are accelerated using a high voltage (typically between 5 kV and 20 kV) to form a beam of sufficient energy to give an adequately bright spot on the phosphor. The beam can be deflected electrostatically by two sets of parallel metal plates, one to deflect the beam horizontally and one to deflect vertically. To achieve a display that mimics the voltage waveform of an input signal, the electron beam must be swept horizontally at a steady rate and deflected vertically in proportion to the instantaneous voltage of the input signal. In addition, the beam must be swept repeatedly along the same path in order to achieve sufficient brightness to be easily seen. This is achieved by the trigger circuit and timebase, which generate a repeating ramp waveform synchronized with the input signal. The operation of the trigger circuit is described further in the next section.

The two input **channels** have variable input attenuators and amplifiers to accommodate a wide range of signal amplitudes. Coarse adjustment of displayed amplitude is normally by means of a rotary switch marked in volts per division (of the **graticule** inscribed on a plastic sheet in front of the CRT). Fine adjustment is also provided by a continuously variable knob with a detent or "click" at one position (usually fully clockwise or anticlockwise) known as the calibrated or CAL position. When the knob is in this position, and provided that the oscilloscope has been

The term **cathode ray** is archaic, but still used in this context.

Electron guns, acceleration, and electrostatic deflection are explained in more detail by Compton (1990).

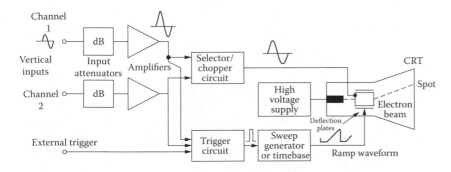

Figure 6.8   Simplified block diagram of a conventional two-channel single-beam oscilloscope.

calibrated by means of internal adjustments against a standard voltage source, readings may be taken from the graticule and converted to voltages using the switch setting in volts per division. Absolute (as opposed to relative) voltage measurements must be made with the variable knob in the calibrated position. Some more expensive oscilloscopes had a red warning lamp to show when the knob was not in the calibrated position. Typical amplitude ranges are from 5 mV to 20 V per division on a general purpose oscilloscope.

To allow both input signals to be displayed simultaneously, a circuit is provided to select or chop the two signals to make a single signal to deflect the electron beam vertically. This can be done either by selecting one channel per horizontal sweep alternating between the two channels, usually known as *alternate mode*, or by switching rapidly between the two channels as the beam sweeps across the display. This rapid switching is known as **chopping**, and is primarily useful for low frequency signals where the slow sweep speed would otherwise allow the eye to see the separate sweeps of an alternate display. False display of the timing relationship between two signals can occur with alternate sweeps. Chopping avoids this problem but often gives a confusing display at high sweep speeds. Some more expensive analogue oscilloscopes had a dual-beam CRT with two electron beams and deflection systems so that two signals could be displayed in **exact** time relationship to each other without chopping. An important point to realize about an oscilloscope display is that the waveform seen is a time average of multiple sweeps (except when sweeping at a very low rate so that the "spot" can be seen to move across the screen). The averaging arises from the persistence of the human eye (about 1/16 second) and causes the displayed waveform

Random noise averages to zero.

to show less noise than might really be the case. This is soon realized when one uses a digital oscilloscope to capture a single sweep: the displayed waveform may not be as smooth as one would expect from experience with analogue oscilloscopes. Of course, in many cases the visual averaging provided by the eye when using an analogue oscilloscope is an advantage in reducing error in a measurement. Most digital oscilloscopes have some capability to average the signal so that a waveform can be shown without the noise content.

Triggering

To obtain a stable, clear display of the input waveform, it is vital to synchronize the timebase circuit so that the horizontal sweep starts at the same point in the input waveform on each display sweep. This is achieved by the **trigger circuit**, which is able to start the timebase at any time, dependent on either the input waveform or on some other trigger generated externally to the oscilloscope. In the absence of a trigger signal the timebase circuit may operate in **free-running** mode in which a periodic (repeated) ramp waveform is generated. Most oscilloscopes

Some analogue oscilloscopes also had a video trigger mode to allow display of analogue composite video (baseband television) signals.

offer the user a choice of several triggering **modes**. NORMAL mode triggers the timebase sweep only when a trigger signal is present and crosses a threshold level in a given direction (positive-going or negative-going). The threshold level can be adjusted manually to obtain the desired display, and usually an indicator lamp shows when the trigger circuit is operating. AUTO triggering mode triggers the timebase in synchronism with the input signal if one is present but otherwise allows the timebase to free-run. In this way, a trace is seen if the input signal is a steady voltage or if the input attenuator is connected to ground for the purpose of adjusting the reference level position, as described in the next section.

An important consideration when displaying two signals simultaneously is the choice of signal to operate the trigger circuit, particularly when the two signals are of different frequency but are related to each other. If one signal is derived by frequency division from the other, for example, the lower frequency signal should be selected as the trigger. If this is not done, the low frequency signal will not be correctly displayed because the trigger circuit will trigger from the higher frequency signal whose cycles occur at various phase angles relative to the lower frequency signal.

Input coupling

The input signal from a front panel socket on an oscilloscope can normally be coupled to the input attenuators directly (DC COUPLING), through a capacitor (AC COUPLING) or to 0 V (GROUND or GND). DC COUPLING is normally the best choice and is essential for making voltage measurements relative to 0 V. To do this, the GROUND position is provided to allow the trace to be aligned against a horizontal graticule line on the face of the CRT, which is chosen to represent 0 V. To align the trace, AUTO triggering is selected so that a trace is displayed even though there is no input signal, and the trace is adjusted using the vertical POSITION control. The coupling switch can then be set to DC to display the input signal and, provided that the vertical POSITION control is not altered, voltages relative to 0 V may be read off relative to the chosen 0 V graticule line. The vertical position adjustment may **drift** over time, particularly if the oscilloscope is not fully "warmed up," and the trace should be checked using the GROUND setting and, if necessary, readjusted immediately before making a measurement or series of measurements.

The AC COUPLING mode is provided to allow display of a small time-varying signal superimposed on a larger d.c. level. The coupling capacitor blocks the d.c. component and passes the a.c. or time-varying component (which need not be sinusoidal). In a transistor amplifier, as shown in the margin for example, the d.c. component would be the bias voltage at the point under examination (4 V), while the a.c. component could be of a few millivolts. The input attenuator would be set to perhaps 1 mV/division to display the a.c. component, but this would cause the bias voltage to saturate the input amplifiers if DC COUPLING were selected, causing the electron beam to be deflected beyond the top or bottom of the CRT face.

The choice of AC or DC COUPLING also occurs with an external trigger signal, and similar considerations apply. On some oscilloscopes, additional settings are provided for rejecting or blocking either low-frequency or high-frequency components.

Many digital oscilloscopes are operated by very similar controls, although they also have menu-driven options displayed on their liquid crystal displays and selected by adjacent buttons. Internally, a digital oscilloscope operates quite differently to an analogue instrument, since the waveform must be digitized by an ADC and stored in memory. The stored **samples** of the input signal are then used to construct the displayed waveform. The principles of triggering are similar, even though the internal implementation is different. Digital oscilloscopes offer more functionality in the form of software-implemented functions such as automatic calculation of period or frequency, and time interval and voltage measurement between on-screen cursors. A digital oscilloscope can also capture a **one-shot** waveform (for example, the voltage output of a power supply during switch-on) that could not be captured or displayed on an analogue instrument.

Parasitic loading — The 10 × probe

Any measuring instrument, the oscilloscope included, imposes a **parasitic load** on the circuit to which it is attached. A typical modern oscilloscope has an input impedance of 1 M Ω in parallel with about 25 pF. (Some oscilloscopes for radio frequency work have 50 Ω inputs for matching to 50 Ω transmission lines.) In many circuits, the loading effect of a 1 MΩ oscilloscope input is negligible, because the current drawn by the 1 MΩ load is much smaller than the current flowing in the circuit being measured. Therefore the voltage in the circuit is not altered significantly by the attachment of the oscilloscope. In some circuits, however, there may be high impedance circuit nodes where the voltage would be significantly altered by the 1 MΩ loading of an oscilloscope.

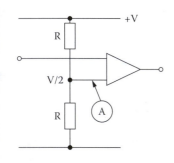

Worked Example 6.3

Consider the comparator circuit shown in the margin. A threshold level of $V/2$ is obtained by a potential divider with two equal resistances, R. By what percentage will the voltage at A be modified by the attachment of a 1 MΩ oscilloscope if R is (a) 100 kΩ; (b) 10 kΩ; and (c) 1 kΩ?

Solution   The voltage, A, will be given by $VR'/(R + R')$, where R' is the equivalent resistance formed by R in parallel with the 1 MΩ oscilloscope input, assuming the input impedance of the comparator to be infinite. The value of R' to the nearest 1 Ω for each of the values of R given will be (a) 90 909 Ω, (b) 9901 Ω, and (c) 999 Ω. Without the oscilloscope, the voltage at A is $V/2$, thus the percentage difference in voltage when the oscilloscope is attached will be

$$\frac{(V/2) - (VR'/(R + R'))}{V/2} \times 100\%$$

or

$$\frac{1 - 2R'}{R + R'} \times 100\%$$

For the values of R given, the percentage differences are

(a) 4.8%, (b) 0.50%, and (c) 0.05%.

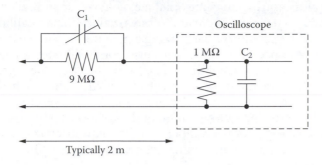

Figure 6.9   Schematic diagram of a 10× probe.

Generalizing from this worked example, if we consider a circuit node where an oscilloscope is to be attached in terms of its Thévenin equivalent circuit as shown in the margin, the Thévenin resistance, R, must be less than 10 kΩ for the 1 MΩ input resistance of the oscilloscope to alter the circuit voltage V_0 by less than 1%. Many circuits will have Thévenin resistances greater than 10 kΩ, and some measure must be taken to reduce the loading effect of the oscilloscope. The circuit of Figure 6.9 shows a very widely used solution, known as a 10× (pronounced "ten times") probe. The 10× probe increases the input resistance of the oscilloscope to 10 MΩ, but reduces the input signal amplitude by a factor of 10 as seen at the oscilloscope input socket. The reduction in amplitude is not a problem, except for small signals, since the oscilloscope input amplifiers can amplify the attenuated signal to compensate. Most oscilloscopes have input voltage range switches marked with two positions or two sets of figures, one for normal use and one for use with a 10× probe. Digital oscilloscopes usually have a menu option to select 1× or 10× inputs, and adjust the displayed values accordingly.

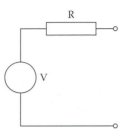

The 10× probe has a further complication. This is that the attenuation of the probe is frequency dependent unless the ratio of its capacitance, C_1, to the input capacitance of the oscilloscope, C_2, is 1:9. This means that 9 MΩ × C_1 = 1 MΩ × C_2, and ensures a flat frequency response. Because C_2 varies from one oscilloscope to another and between two channels on the same oscilloscope, C_1 must be variable and it must be adjusted for a particular oscilloscope input. This is conveniently done by connecting the probe to a low-frequency square wave, typically 1 kHz, which is provided at a front-panel terminal on many oscilloscopes. Once the 1 kHz square wave is displayed, C_1 is adjusted to give a "square" waveform as shown in the margin at B. Waveforms A or C will be obtained when C_1 is too small or too large respectively. The adjustable capacitor, C_1, may be located at the oscilloscope end of the probe on some designs: the principle is the same.

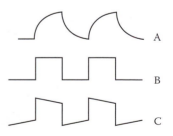

Summary

This chapter has reviewed the fundamental quantities that an electronics engineer may need to measure: voltage, current, frequency, phase shift, time delay, and rise and fall time. Voltage and current may be measured using electromechanical or electronic meters. The potential problem of measuring nonsinusoidal a.c. voltages or currents has been highlighted — most meters read the r.m.s. value, but are calibrated for sinusoids. The operation of the moving coil ammeter and the dual-slope integrator have been described. The principles of frequency and time measurement using counter/timer instruments have been presented, including the need to ensure correct threshold settings. The oscilloscope has been introduced as the most widely used electronic instrument, and its principles of operation and use have been presented. The correct use of triggering and input coupling have been discussed. The problem of parasitic circuit loading and its solution, the 10× probe, have been introduced.

Problems

6.1 A 1 mA f.s.d. moving coil meter with a coil resistance of 75 Ω ± 1% is to be shunted with a suitable resistor to give an f.s.d. of 100 mA. (a) Calculate the required resistor value. (b) Assuming the shunt has a tolerance of ±1%, what are the worst case errors in the meter reading when measuring a 100 mA current? Are these errors significant?

6.2 An oscilloscope with a 1 MΩ input and a 10× probe is to be used to measure the voltage at a point, P, in a circuit with a Thévenin equivalent resistance, R. What is the highest value of R allowable if the loading effect of the oscilloscope is to alter the voltage of P by less than 2% (a) with and (b) without the 10× probe?

Heat management

7

Objectives

☐ To introduce the physical principles of heat and common solutions to heat problems in electronic systems.

☐ To review heat-transfer mechanisms in the context of electronic systems.

☐ To introduce heat sink technology and simple heat sink calculations.

☐ To discuss briefly some heat-removal methods used in high-density electronic systems.

Heat is generated in all operating electronic systems as a consequence of the flow of electric current. Heat is energy (in the form of molecular or atomic vibrations) and is measured in the same units (joules) as any other form of energy. Rate of generation or transfer of heat is therefore expressed in joules per second, or watts.

Heat generation in a conductor carrying electric current arises from interactions between the charge carriers and the atoms within the conductor, resulting in the transfer of some energy from the charge carriers to the atoms of the conductor. At circuit level, the concept of electrical resistance is used to explain the heating effect of an electric current. Heat can also be generated in parts of an electronic system where current flow is either undesired or unexpected. Examples include: eddy currents circulating in the magnetic core of a transformer or inductor, causing heating of the core; heating in the walls of a waveguide due to induced currents caused by the propagation of electromagnetic waves through the guide; and heat generated in the dielectric of a capacitor while charging and discharging.

The most obvious and extreme effect of heat generation in electronic components is thermal destruction: most electronics engineers have accidentally destroyed components during testing of breadboard or prototype circuits by applying a power-supply voltage that is too high or by inadvertently causing a short circuit, resulting in excessive heating and consequent destruction.

Rated operating temperatures of electronic components can be surprisingly high. Military-grade integrated circuits and transistors, for example, can be operated at ambient temperatures of 125°C, while some power resistors can operate at over 150°C. Despite this apparent robustness, most components in commercial and military designs are operated at much lower temperatures by conservative choice of device rating or by taking special measures to remove heat. The reason for this is that reliability decreases with increasing temperature because higher temperatures accelerate many physical and chemical processes such as diffusion, material creep, and corrosion that can contribute to component deterioration and eventual failure. Repeated temperature changes

Heat generation in a resistance, R, due to an imposed voltage, V, or a current, I, is given by V^2/R or I^2R or VI. These expressions are valid for steady voltages and currents. In a.c. circuits, phasor multiplication and division must be used.

Dielectric losses in capacitors were discussed in Chapter 5.

If a resistor is required to dissipate ⅛ W in a circuit application, a conservative choice of component might be a ¼ W or ½ W resistor, either of which would operate at a lower temperature than a ⅛ W device, while dissipating ⅛ W.

Reliability is defined in Chapter 9, and the effect of temperature on reliability is discussed in Chapter 10.

(thermal cycling) have an even more drastic effect on component life, because differential expansion within a component induces mechanical stress, which can lead to cracking of materials or separation of bonded joints.

Heat transfer

For more information on heat transfer, see Wong (1977) or Long (1999).

There are three main physical mechanisms by which heat can be transferred: conduction, convection, and radiation. The first two of these are the most important in electronic engineering, although radiation can contribute to heat dissipation from a blackened heat sink. One important exception to this general statement, however, is in satellite design, where radiation is the only mechanism by which heat can be lost ultimately from the craft.

Thermal conduction

Both electrical and thermal conductivity depend upon electron mobility, and good electrical conductors (metals) are also good thermal conductors for this reason.

Conduction is important in solid materials such as silicon, aluminium, copper, and plastic. Good electrical conductors are also good thermal conductors. Conversely, electrical insulators are poor conductors of heat. Thermal conduction is the first process involved in removing heat from the interior of an electronic component.

The ability of a material to conduct heat is quantified by the thermal conductivity of the material, k, expressed in W m^{-1} °C^{-1} and defined by the expression

Thermal conductivity is properly quoted in W m^{-1}K^{-1}, but since temperature difference rather than absolute temperature is used, °C has been used here. This is also consistent with the use of °C in stating thermal resistance values.

$$\frac{dQ}{dt} = -kA\frac{d\theta}{dx} \tag{7.1}$$

where dQ/dt is the rate of heat flow (W) through a cross-sectional area, A, of the material (m^2), and $d\theta/dx$ is the temperature gradient (°C m^{-1}) normal to the area.

Convection

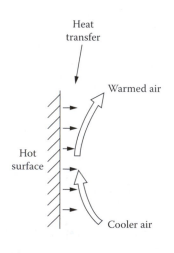

Convection is an important heat transfer mechanism in fluids (gases and liquids), and is the means by which heat is ultimately removed from most electronic systems, usually by transfer of heat to the surrounding air. Convective heat transfer from a hot surface to the nearby air is shown in the margin. The air immediately adjacent to the hot surface becomes warmed by conduction and decreases in density as a result of the increase in temperature. The warmed air is then displaced by cooler air of higher density. The displaced warm air, of course, carries heat away with it. Convective heat transfer can be enhanced by increasing the area of the hot surface (by adding fins), by placing the hot surface vertically, and by forcing cooling air across the surface (forced convection). Convective heat transfer is governed by the same form of expression as was given above for conduction.

This equation is known as Newton's law of cooling.

$$\frac{dQ}{dt} = hA\Delta\theta \tag{7.2}$$

where dQ/dt is the rate of heat loss (W) from an area, A (m^2). $\Delta\theta$ is the temperature difference between the hot surface and the surrounding fluid (°C). h is then a convective heat transfer coefficient (W m^{-2} °C^{-1}). Unfortunately, the coefficient h depends on temperature and is not easily calculated. As we shall see below, however, for simple heat-removal design problems, we can use the concept of thermal resistance.

Radiation

Hot surfaces, even in a vacuum, can lose heat by emission of electromagnetic radiation. At the temperatures encountered in electronic systems (say, –50°C to 150°C), radiated energy will be in the microwave and infrared regions of the electromagnetic spectrum. Heat loss by radiation is governed by the expression

$$\frac{dQ}{dt} = \sigma\varepsilon A(\theta_s^4 - \theta_A^4) \tag{7.3}$$

This expression implies assumptions about the natures of the emitting surface and the surroundings. See Wong (1977) for more information.

where dQ/dt is the rate of heat loss (W) from an area A, (m^2); σ is the Stefan-Boltzmann constant (5.67 × 10^{-8} W m^{-2} K^{-4}); θ_s is the temperature of the emitting surface (K); and θ_A is the temperature of the surroundings (K). ε is the emissivity of the hot surface (dimensionless and ≤ 1).

Thermal resistance

Practical heat calculations are simplified by using the concept of thermal resistance, which has the units of °C W^{-1}. For a general heat transfer problem,

$$\Delta\theta = \frac{dQ}{dt} R_\theta \tag{7.4}$$

where $\Delta\theta$ is a temperature difference (°C), dQ/dt is heat transfer rate (W), and R_θ is a thermal resistance (°C W^{-1}). This equation is analogous to Ohm's law with temperature replacing voltage, heat transfer rate replacing current, and thermal resistance replacing electrical resistance. The thermal resistance can take account of conduction, convection, and radiation effects provided we use the thermal resistance value only as an approximation and only over a limited range of temperatures.

In most applications, the thermal resistance values needed can be found in manufacturer's data sheets, and there is no need to perform calculations from first principles.

Exercise 7.1

Calculate the thermal resistance between the end faces of a 10 mm × 10 mm aluminium alloy bar of length 100 mm. The thermal conductivity, k, of the alloy is 170 W m^{-1} °C^{-1}. Ignore heat loss from the sides of the bar.

(*Answer*: about 6°C W^{-1}.)

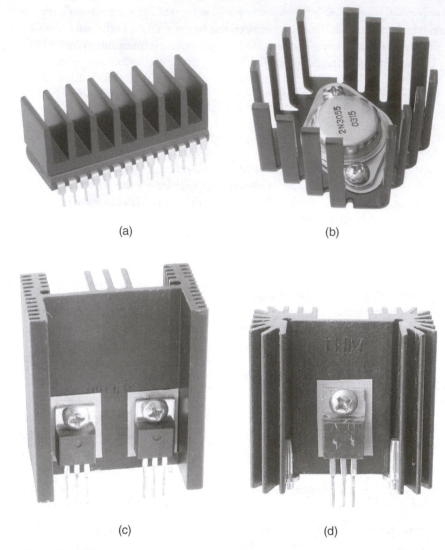

(a)

(b)

(c)

(d)

Figure 7.1 Heat sink designs: (a) 25°C W^{-1}, (b) 5.1°C W^{-1}, (c) 4.2°C W^{-1}, and (d) 4.4°C W^{-1}. (Courtesy of Aavid Thermalloy.)

Heat sinking

The term *heat sink*, meaning a device for conducting heat from a power semiconductor and dissipating that heat to the surroundings, is poor terminology: the true heat sink is the surrounding environment.

Heat sinks are used to conduct heat from power semiconductors and resistors, and to dissipate this heat to the surroundings. The main heat-dissipation mechanism is convective heat transfer to the surrounding air, although radiation can also contribute to the heat loss. Many heat sinks have a blackened surface finish (which adds negligible cost to the heat sink) to enhance radiated energy loss. For higher-power dissipations, fan-assisted removal of heat from heat sinks is common, particularly for processors in desktop and laptop computers. Small power devices dissipating up to about 2 W may be cooled by clip-on heat dissipators. Figure 7.1 shows a selection of heat sinks and heat dissipators. All of these devices are made from aluminium alloy because of its

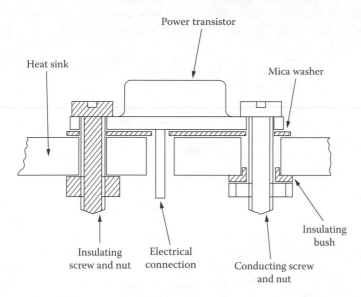

Figure 7.2   Mounting arrangements for a power transistor showing alternative fixing arrangements for electrical insulation between the transistor and the heat sink. (Normally both screws would be insulated in the same way.)

high thermal conductivity and the ease with which it can be extruded to form the complex finned shapes needed for effective convective heat transfer.

Designers of power semiconductors pay careful attention to the thermal design of their packages to ensure the lowest possible thermal resistance between the power-dissipating regions of the device and the outer mounting surface of the case. In many instances, the metal case serves as one of the electrical terminals. Quite often the case terminal is not at 0 V potential, and the power device has to be insulated from the heat sink. Unfortunately, electrical insulators are also good thermal insulators and only a thin layer of insulator can be allowed if the thermal resistance between the case and the heat sink is to be kept low. Thin washers of mica (a naturally occurring laminated mineral) are often used and are available to suit all types of power semiconductor package. Insulating screws (of nylon) or insulated bushes for use with metal screws are needed to fasten a power device to a heat sink. Figure 7.2 shows a typical power transistor of the type shown in Figure 7.1b mounted on a heat sink and illustrates the alternative fastening methods.

Mica is also used in silvered mica capacitors, as discussed in Chapter 5.

To improve heat transfer from the case of a power semiconductor through a mica washer to a heat sink, a heat-conducting compound such as silicone grease is smeared onto both sides of the washer before assembly. The grease fills voids left between the metal-to-washer interfaces that would otherwise be filled with air, which is a poor thermal conductor and would increase the overall thermal resistance.

Some device data sheets do not state thermal resistance directly: instead, they may include a power-dissipation derating curve similar to that shown in Figure 7.3 or an equivalent statement in words. The thermal resistance of the device is the reciprocal of the slope of the derating curve (the thermal resistance is, of course, always positive).

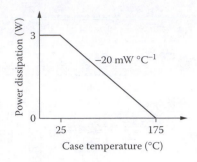

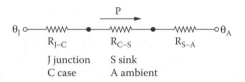

Figure 7.3 A typical power-dissipation derating curve for a small silicon transistor.

Figure 7.4 Electrical analogue circuit for a power device on a heat sink (steady state).

Steady-state heat sink calculations

The simplest heat sink situation to handle is steady-state power dissipation with one power device on a heat sink. (The same calculation applies for a small power device with a clip-on heat dissipator except that there would be no mica washer.) Figure 7.4 shows the electrical analogue of the thermal path. The variables in this arrangement are: the power dissipation, P (W); the thermal resistances, R (°C W^{-1}), which are all in series; and the two temperatures, θ_A (the ambient temperature of the surroundings) and θ_J (the internal junction temperature of the power device). These quantities are related by the thermal analogue of Ohm's law, which is

$$\theta_J - \theta_A = P(R_{J-C} + R_{C-S} + R_{S-A}) \tag{7.5}$$

Of these quantities, R_{J-C} and R_{C-S} are often known from the choice of power device, while various combinations of the remaining values are to be decided. As with many design problems, the final choice of heat sink or allowable power dissipation may be arrived at after several iterations.

Strictly, the term *junction temperature* applies only to bipolar devices, but it is also used conventionally for unipolar devices such as power MOSFETs, which do not depend on a p–n junction, and for digital circuits.

Worked Example 7.1

A voltage regulator integrated circuit (IC) dissipating 25 W is to be mounted on a heat sink such that the junction (internal) temperature of the IC is to be limited to 125°C. What is the maximum allowable thermal resistance of the heat sink if the thermal resistance of the regulator (junction–case) is 1.7°C W^{-1}, the thermal resistance of the mica washer and silicone grease is 0.3°C W^{-1}, and the ambient temperature is 25°C?

Solution   Total thermal resistance allowable = (125°C − 25°C)/25 W = 4°C W^{-1}. Thermal resistance, junction–case, and mica washer is 1.7 + 0.3 = 2°C W^{-1}. Therefore the maximum allowable thermal resistance of the heat sink is 4 − 2 = 2°C W^{-1}.

Exercise 7.2

For the conditions of Worked Example 7.1, if the heat sink actually used had a thermal resistance of 1.5°C W^{-1}, what would the junction temperature of the IC be?

(*Answer*: 113°C.)

Transient or dynamic heat sink calculations

We have seen in the previous section how to select a heat sink to dissipate heat from a power device with a steady power dissipation. Worked Example 7.1 has suggested an application where such a calculation might be used. In other applications, however, power dissipation may not be steady: it may fluctuate unpredictably (as in an audio power amplifier), or it may be pulsed with a regular and predictable frequency and waveform. Sometimes, it may be necessary to design for a worst-case maximum continuous power dissipation using the steady-state methods described above. Otherwise, it may be possible to use a smaller, cheaper heat sink by taking account of the pulsed nature of the power dissipation. Figure 7.5 illustrates why this is so: the power device itself and the heat sink possess thermal capacitance (analogous to electrical capacitance). If the power dissipation is intermittent, the thermal capacitances are able to absorb heat during the time that power is being dissipated in the semiconductor device and release this heat through the thermal resistances during the remaining time. In reality, the thermal capacitances are distributed physically throughout the power device and heat sink and are not lumped as shown in the figure. This means that an analytical approach to heat sink calculations under pulse conditions is not feasible: instead, we use empirical data provided by the device manufacturers from experiments, tests, and possibly finite-element computer modelling.

Figure 7.6 shows the form of the data provided by the manufacturers of a power device operating under pulse conditions. The quantity δ

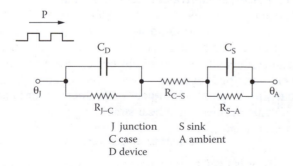

Figure 7.5   Electrical analogue circuit for a power device on a heat sink (transient power dissipation).

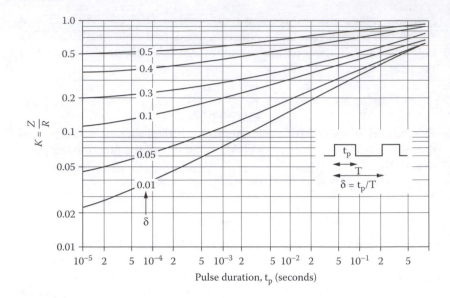

Figure 7.6 Typical transient thermal impedance curves for a power semiconductor device.

is the duty cycle of the device, the ratio of on-time, t_p, to the period of the pulse waveform, T; K is the ratio of transient thermal impedance, Z, to steady-state thermal resistance, R. With a pulsed power dissipation, Equation 7.5 becomes

$$\theta_J - \theta_A = P_{max}(KR_{J-C} + \delta R_{C-S} + \delta R_{S-A}) \tag{7.6}$$

The thermal resistance from junction to case has been multiplied by the factor K, found from the manufacturer's data. Since K is less than 1, the effective thermal impedance from the junction to the case is lower, reflecting the fact that the mass of the device can absorb heat temporarily during the power pulses (in its thermal capacitance). The thermal resistances of the mica washer and the heat sink have simply been multiplied by the duty cycle of the pulse waveform, δ, to allow for the average power dissipation. (This is conservative: in fact, a practical heat sink can have substantial thermal capacitance.)

If the actual operating power waveform of the device does not consist of rectangular pulses, an equivalent rectangular pulse waveform must be used to find δ and K. Clearly, the areas under the actual and equivalent waveform pulses must be the same so that they contain equal energies. One can keep either the peak amplitudes equal or the pulse durations equal: the choice depends on the actual pulse waveform and is largely a matter of engineering judgement. (One could, of course, calculate the required heat sink thermal resistance for both cases and use the more conservative value in selecting a heat sink.)

Worked Example 7.2

A power transistor is used to switch a resistive load with a duty cycle of 30% and a pulse duration of 20 ms. When conducting, the device dissipates 25 W. The thermal resistance, junction–case, is 1.5°C W^{-1}, and a mica washer adds 0.3°C W^{-1}. What is the maximum thermal resistance of the heat sink required for an ambient temperature of 50°C and a junction temperature of 150°C? Use the transient thermal impedance curves of Figure 7.6.

Solution   From Figure 7.6, $K = 0.4$ ($\delta = 0.3$, $t_p = 2 \times 10^{-2}$ s)

Using Equation 7.6

$$150 - 50 = 25(0.4 \times 1.5 + 0.3 \times 0.3 + 0.3 \times R_{S-A})$$
$$R_{S-A} = 11°C\ W^{-1}$$

Exercise 7.3

If, in Worked Example 7.2, the device dissipation was assumed to be a steady 25 W (a conservative assumption), what would be the maximum thermal resistance of the heat sink?

(*Answer*: 2.2°C W^{-1}.)

Forced cooling

The possibility of enhancing convective heat loss from a heat sink by forcing air across the surface of the heat sink has already been mentioned. Forced air cooling can also be used to remove heat from printed circuit boards (PCBs) mounted in an equipment cabinet or rack, thereby reducing the operating temperature of components mounted on the PCBs and improving their long-term reliability. Figure 7.7 illustrates a typical application of forced air cooling. A small axial fan of about 100 mm diameter draws air through a dust filter and forces the air among a number of PCBs. The air picks up heat from the components on the PCBs and is exhausted through louvres or slots. Electric fans for this type of application are readily available in a range of sizes and operable from either a mains supply or a low-voltage d.c. supply. A volumetric air flow rate of around 100 m³hour^{-1} can be produced by a fan of 100 mm diameter. Vertical mounting of PCBs can be an advantage for convective cooling, whether forced or not.

Volumetric flow rates may still be quoted in cubic feet per minute (CFM) by some manufacturers, especially in the USA. 1 CFM $\cong$ 1.70 m³ hour^{-1}.

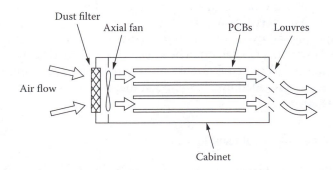

Figure 7.7   A typical arrangement for force-cooling PCBs.

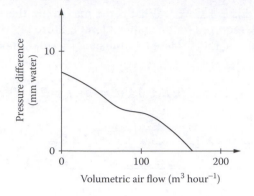

Figure 7.8 Typical air flow characteristic for a small axial fan of about 100 mm in diameter.

Worked Example 7.3

Derive an expression for the volumetric air flow rate required to limit the internal temperature of an enclosure to $\theta°C$ above ambient when the power dissipation inside the enclosure is P.

Solution When steady-state conditions have been attained, the air flowing through the enclosure must remove P watts, or $3600 \times P$ joules hour^{-1}. If the specific heat capacity of air is c (J kg^{-1} °C^{-1}), then the mass flow rate needed is $3600P/c\theta$ (kg hour^{-1}). Dividing by the density, ρ (kg m^{-3}) we obtain the volumetric flow rate required:

$$\frac{3600P}{\rho c \theta} \text{(m}^3 \text{ hour}^{-1})$$

The air flow rate produced by a fan depends upon the resistance to air flow presented by filters, louvres, and PCBs. Each resistance causes a pressure drop (analogous to voltage drop across an electrical resistance) and the total pressure drop must be overcome by the fan. Fan manufacturers state the performance of their products by means of a characteristic showing pressure difference against volumetric flow rate (Figure 7.8). The characteristic of a dust filter, or other air flow resistance, would be plotted in the same way, showing pressure drop versus flow rate. The flow rate achievable by a given fan with given air flow resistance characteristics for the filter and enclosure can be found by superimposing the two sets of characteristics as shown in Figure 7.9. In practice, however, a fan can often be selected by calculating the required air flow rate as in Worked Example 7.3 and then applying empirical correction factors provided by filter and fan manufacturers to allow for air flow resistance.

Advanced heat-removal techniques

In many modern electronic engineering applications, the techniques so far described are sufficient to deal with heat-removal problems. Most

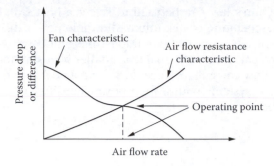

Figure 7.9 Graphical determination of the air flow rate from fan and air flow resistance characteristics.

printed circuit boards can be cooled by natural convection or gentle fan-assisted convection with no special design measures necessary on the PCB. Indeed, some digital electronic systems implemented in complementary metal-oxide semiconductor (CMOS) logic consume so little power (and therefore dissipate so little heat) that they can be housed in a totally sealed enclosure. The reason that these systems present no special heat-removal problem is that they have a low power density (power dissipation per unit volume or per unit of PCB area). In applications where space is at a premium, such as on board an aircraft, designers may be forced to increase packing density (and therefore power density).

The main problem with heat removal from a high-density system is the low thermal conductivity of PCB laminates. Multilayer boards with internal power and ground planes have a higher thermal conductance than single- or double-sided boards, but still pose a problem in ultimate heat removal from the board. One possible solution to this problem is a metal heat ladder bonded to the PCB. This requires the components on the board to be positioned in the areas not occupied by the heat ladder. Special provision is made at the board edge to remove heat from the heat ladder for conduction to the outside surface of the equipment.

In very high-speed mainframe supercomputers, components have to be placed physically close together because of the finite speed of light, resulting in power densities so high that liquid (usually water) cooling has to be used. The physical design of such systems then becomes dominated by the coolant system rather than the PCB.

Summary

Heat is an unavoidable consequence of the flow of electric current and is therefore produced in all electronic equipment wherever electric current flows or circulates. Heat transfer can take place by three physical processes: conduction, convection, and radiation. Hot electronic components are less reliable than cooler ones.

Practical heat calculations use the notion of thermal resistance and an analogy with electrical resistance in which temperature difference replaces voltage and heat flow rate replaces electric current.

Power transistors and integrated circuits are kept cool by attaching them to heat sinks. For steady-state power dissipation, the thermal analogue of Ohm's law can be used to select a heat sink or calculate the internal temperature or maximum allowable power dissipation of a power device. Under transient conditions, thermal capacitance can be included in the calculations to enable a smaller heat sink to be used.

Forced air flow can enhance heat loss from a heat sink or PCB.

In electronic systems with a high power density, special measures may be needed to remove heat.

Problems

7.1 Two transistors, each dissipating 15 W and having a thermal resistance (junction–case) of 1.5°C W^{-1}, are to be mounted together on the same heat sink. The junction temperatures are to be limited to 60°C above ambient temperature. Assume the thermal resistance of mica washers and thermal grease to be 0.4°C W^{-1}. What is the maximum allowable thermal resistance of the heat sink?

7.2 A designer wishes to use a small transistor dissipating 1.2 W. The manufacturer's data sheet does not state a thermal resistance value but gives the power-dissipation derating curve shown in Figure 7.3. A heat dissipator with a thermal resistance of 48°C W^{-1} is available to fit the transistor. What is (a) the maximum safe dissipation with the dissipator, and (b) the case temperature at a power dissipation of 1.2 W? Assume an ambient temperature of 25°C.

7.3 A cabinet containing circuitry dissipating 150 W is to be cooled by a ventilating fan. The ambient temperature inside the cabinet is to be limited to 50°C for an air inlet temperature of 30°C. What is the minimum air flow rate required? The specific heat capacity of air at 30°C is about 1000 J $kg^{-1}°C^{-1}$, and the density of air at 30°C and normal atmospheric pressure is about 1.3 kg m^{-3}.

Parasitic electrical and electromagnetic effects

8

Objectives

- [ ] To introduce the problem of parasitic circuit elements and the limitations of lumped-parameter circuit models.
- [ ] To review electromagnetic induction effects.
- [ ] To introduce the problem of electromagnetic interference and the subject of electromagnetic compatibility.
- [ ] To present studies of common problems in real circuits and examples of good practice in circuit and system design.

The behaviour of many electronic circuits can be influenced by parasitic electrical and electromagnetic effects that are incidental to the intended properties of the circuit. Troublesome symptoms in analogue systems include crosstalk or unwanted coupling of a signal from its intended pathway to some other pathway; instability or spurious oscillation; pickup of unwanted signals, especially at mains frequencies; and microphony or sensitivity to mechanical disturbance or vibration. Digital systems can be prone to data errors and spurious states caused by crosstalk and switching transients. Both types of systems can be susceptible to electrical or electromagnetic interference (EMI), which is often caused by other electrical or electronic equipment, and can themselves be sources of such interference.

Many of these problems can be controlled, often by the application of quite simple design measures and good practice. In order to understand how these design measures work, we must first examine mechanisms that allow electromagnetic energy to stray from its intended path.

Parasitic circuit elements

Any electronic circuit contains parasitic or "stray" circuit elements. Some of these are inherent in the circuit components, as discussed in Chapter 5, while others are properties of the physical layout and surroundings of the circuit. Sometimes the parasitic elements have very little effect on circuit performance and can be ignored. In other cases, however, the ultimate level of performance achievable from a circuit can be determined by parasitic effects. The three types of passive parasitic circuit elements will now be considered.

Parasitic resistance is not often a problem since the resistance of a wire or printed circuit board (PCB) track is usually negligible compared to the impedances of a circuit, and the insulation resistance between wires or tracks is so much higher than the circuit impedance that voltage drops and leakage currents have negligible effect. Nevertheless, it

Parasitic circuit elements are not restricted to passive devices. In integrated circuits (ICs), parasitic active elements such as transistors are often inherent in a particular IC process.

Important exceptions include the resistance of power-supply rails and internal series resistance in some integrated circuits.

is as well to remember that voltage drops and leakage currents do exist and that insulation can break down at sufficiently high voltages.

Parasitic reactance is a much more common problem in electronic circuits. It can be understood in terms of parasitic capacitance and inductance, or in terms of the equivalent electric and magnetic fields generated by a circuit. Both points of view are valid and useful, but as we shall see, the field concept is more general and gives a better understanding of control measures used to reduce the magnitude of the reactances. A parasitic capacitance exists between any pair of conductors and can couple energy from one conductor to the other. Similarly parasitic inductance can exist either as self-inductance, which can be important in power-supply circuits, particularly in digital systems, or because any pair of circuits has a mutual inductance, which can couple energy from one circuit to another.

Parasitic capacitance

The capacitance between two conductors can be defined as $C = Q/V$, where the presence of equal and opposite charges of magnitude, Q, on the conductors results in a potential difference, V, between them. An electric field exists between any two oppositely charged conductors, and the line integral of this field, $\int \mathbf{E} . \mathbf{dl}$, along any path between the conductors is equal to the potential difference, V. We can associate capacitance, therefore, with the electric field between conductors caused by the presence of charge on the conductors. The capacitance between two conductors can be modified by altering the electric field distribution in some way. Normally, we wish to reduce parasitic capacitance, and several ways of doing so are discussed later.

Before going any further, it is useful to have some feel for the magnitude of a parasitic capacitance. We could consider two wires and calculate the capacitance between them, but there is a much easier approach. We can look in an electronic components catalogue to find the value of the smallest capacitor we can buy. This turns out to be about $0.5 - 1$ pF. (0.5 pF surface mount capacitors exist, but for convenience we use 1 pF as the smallest practical value.) Suppose we were to create a 1 pF capacitor using the two sides of a PCB as the plates. What area of copper would be needed, neglecting fringing effects at the edges of the plates?

Strictly, only a closed circuit can have self-inductance since there must always be a return path for current. Later in this chapter, it is explained that the self-inductance of a circuit is reduced if the out and return conductors are kept close together.

Both Compton (1990) and Carter (1992) discussed calculation of stray capacitance.

Worked Example 8.1

A parallel plate capacitor has a capacitance, $C = \varepsilon_o \varepsilon_r A/d$. What area, A, is required to form a 1 pF capacitor if d, the plate separation, is 1.6 mm (a typical PCB thickness) and ε_r (for PCB laminate) is about 6?

Solution   To a good approximation for this purpose, ε_o is 9 pF m^{-1}. The value of $C/\varepsilon_o \varepsilon_r$ is thus $\frac{1}{54}$, and A is $1.6 \times 10^{-3}/54$ or about 30 mm^2. If the plates are square, they will be about 5 mm $\times$ 5 mm.

We can see from this why capacitors much smaller than 1 pF are not commonly available and also that quite small areas of copper on a PCB have a capacitance of this order. Looking again in a catalogue, we can

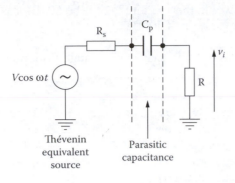

Figure 8.1   Parasitic capacitive coupling between a source and a load.

find that an RG58C/U coaxial cable has a capacitance of 100 pF m^{-1} or 1 pF cm^{-1}, that a nonscreened twisted pair cable has a capacitance of 56 pF m^{-1}, and that a 300 Ω balanced feeder cable that has two straight conductors spaced about 10 mm apart by a plastic web has a capacitance of 13 pF m^{-1}. We could therefore make a reasonable estimate of the capacitance between two PCB tracks or wires a few millimetres apart at, say, 10 to 20 pF m^{-1}, without needing to calculate from first principles.

Because the impedance of a capacitor is inversely proportional to frequency, parasitic capacitances have a greater effect in high-frequency circuits and where high-frequency interference is present. Parasitic capacitances tend to couple high-frequency signals and interference into high-impedance circuits, as shown by the equivalent circuit in Figure 8.1. The Thévenin equivalent source network represents a circuit node where a high-frequency signal is present, coupled through a parasitic capacitance, C_p, to a circuit node with a high impedance to ground, represented by R. v_i is the resulting interference or unwanted signal present across R. This circuit is a potential divider, so that v_i is given by

$$v_i = \frac{R}{R + R_S + 1/j\omega C_p} V \exp(j\omega t)$$

and the magnitude of v_i is

$$|v_i| = \frac{VR}{\sqrt{((R + R_s)^2 + (1/\omega^2 C_p^2))}} \tag{8.1}$$

The sinusoidal voltage, $V \cos \omega t$, is represented by the real part of the complex exponential, $V \exp(j\omega t)$. Equation 8.1 is obtained by taking the modulus of the complex quantity, v_i. For more on complex numbers, see Szymanski (1989).

We want the interference voltage, v_i, to be as small as possible, which means that the denominator in Equation 8.1 must be much greater than R.

Worked Example 8.2

Assume that R in Figure 8.1 represents the input impedance of a moderately high-impedance amplifier, say $R = 100$ kΩ. Let C_p take a typical "stray" value of 0.1 pF. If the Thévenin source network represents an interference source such as a nearby power-supply track carrying high-frequency interference, then the value of R_s may be neglected. At

125

what frequency would the magnitude of v_i be 60 dB less than V? (60 dB represents a voltage ratio of 1000.)

Solution   As $R_s \ll R$, we may approximate Equation 8.1 to give

$$\frac{|v_i|}{V} = \frac{R}{\sqrt{(R^2 + (1/\omega^2 C_p^2)})} = 10^{-3}$$

or

$$\frac{1}{\sqrt{(1 + (1/\omega^2 C_p^2 R^2))}} = 10^{-3}$$

Notice that the frequency is in the audio band. Such effects are not only important at high frequencies.

which gives approximately $\omega C_p R = 10^{-3}$; hence, $\omega = 10^{-3}/RC_p$. Since RC_p = $10^5 \times 10^{-13} = 10^{-8}$, $\omega = 10^5$ rad s^{-1} or 16 kHz.

We can conclude from this example that any high-impedance node in a circuit is highly prone to pick up interference from a nearby source by capacitive coupling. We must therefore be careful when designing the physical layout of the circuit to minimize the parasitic capacitance between any likely interference sources and any high-impedance nodes. We can do this in several ways, the most obvious of which is to keep interference sources well away from the sensitive node, since increased separation reduces the parasitic capacitance. There are two much more effective ways of reducing a parasitic capacitance between two nodes, shown in Figure 8.2. One is to interpose an earthed conducting screen between the two nodes, and the other is to place the two nodes close to an earthed surface or ground plane. The screen or ground plane alters the electric field distribution between the conductors and reduces the field strength near to one conductor due to charge on the other. It also

Carter (1992) discussed the calculation of capacitances using finite element methods on a computer. The electric field distribution can also be mapped by these methods.

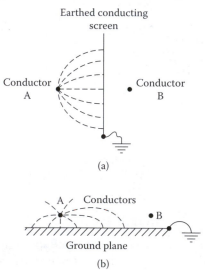

Figure 8.2   Two methods of reducing the capacitance between a pair of conductors: (a) an earthed screen and (b) a ground plane.

increases the capacitance to ground of each conductor. This is an inevitable price to be paid for the reduction in mutual capacitance.

The principle illustrated in Figure 8.2a can be extended so that the screen totally surrounds one of the conductors. This is known as a Faraday cage, and it relies on the well-known result in electrostatics that there can be no electric field inside a conductor, and thus no electric field inside a region enclosed by conductor due to a field outside the enclosure. Note carefully the word *electrostatics*: we are talking here only about static electric fields. In practical circuits, charges move and generate electromagnetic fields. We can begin to look at the problems caused by these by discussing inductance.

The Faraday cage is widely used in screened rooms for electromagnetic testing (discussed in Chapter 10). It is also applied when surrounding the front end of radio receivers, where weak radio-frequency (r.f.) signals are being amplified, to prevent pickup from powerful local sources within the receiver.

Parasitic inductance

Parasitic inductance occurs in two forms. One is the self-inductance of a circuit such as a power-supply rail, load, and return; the other is the mutual inductance between two circuits that can couple unwanted energy from one circuit to another. As we shall see, self-inductance and mutual inductance are closely related, so measures to reduce one can also reduce the other.

Figure 8.3 shows two circuits where the magnetic flux generated by loop 1 links loop 2 and vice versa. Self-inductance and mutual inductance are associated with the magnetic field around a conductor generated by a current. The mutual inductance, M, between two loops can be expressed in terms of the self-inductances of the loops L_1 and L_2 as $M = k\sqrt{(L_1 L_2)}$, where k is a dimensionless factor between 0 and 1 that depends on the proportion of the flux generated by one loop that links the other loop. We cannot easily get a feel for the magnitude of "stray" inductances as we did for stray capacitances, but as we shall see later, we can find the self-inductance per unit length of a coaxial or twisted-pair cable. For the RG58C/U cable, this turns out to be 250 nH m^{-1}; and for the 300 Ω balanced feeder cable with two conductors spaced about 10 mm apart,

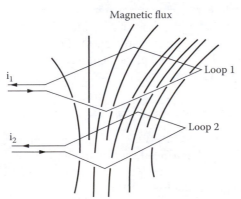

Figure 8.3   Magnetic flux linking two circuit loops.

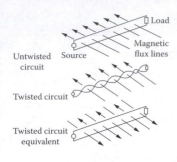

Untwisted
circuit

Twisted circuit

Twisted circuit
equivalent

As we shall see later, the
self-inductance of a circuit is
also reduced by running the
conductors close together, and
this is also desirable.

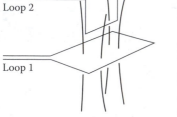

Magnetic flux generated by loop 1.

$1~\mu$H m^{-1}. We can say therefore that the mutual inductance between two pairs of wires running parallel and a few millimeters apart is of the order of 100 nH m^{-1} to $1~\mu$H m^{-1}. To reduce the mutual inductance between circuits and therefore minimize electromagnetic interference problems, we need to reduce the flux linkage between the circuits. This can be done in several ways. In some situations the flux generated by loop 1 is outside our control. It might be the 50/60 Hz flux generated by nearby mains supply wiring. The flux passing through loop 2, however, can be reduced by reducing the area of loop 2. If the wires are twisted together, not only is the area of the loop reduced, but also the direction of magnetic flux lines passing between the two wires alternates with each twist. The effect is as if the flux lines had been made to alternate, cancelling out over the length of the circuit. The flux generated by loop 1 can be reduced by the same method. Wiring to or from a transformer or power supply should therefore be designed so that both conductors are close together throughout their length.

There are three other methods of reducing inductive coupling between circuits, two of which are analogous to those used to reduce capacitive coupling, whereas the other has no capacitive analogue.

As with capacitive coupling, increasing the physical separation between two circuits reduces inductive coupling. Quite often, when designing an electronic system, it can be quite easy to route power-supply wires carrying heavy current away from sensitive circuits. Second, in the same way that an electric screen could be used to reduce coupling by an electric field, a magnetic shield can be used to reduce coupling by a magnetic field. In the electric case, the screen must be of high conductivity, which is easily achieved with copper. In the magnetic case, however, high permeability is needed. A magnetic shielding alloy, such as mumetal, can be used, but it is not as effective against a magnetic field as copper is against an electric field. Magnetic shields are also less effective at very low frequencies.

The third technique mentioned above, which has no analogue in capacitive coupling, is to position loop 2 relative to loop 1 so that the lines of flux generated by loop 1 are parallel to the plane of loop 2.

One final step that we might be able to take to reduce inductive coupling is unconnected with the actual mutual inductance between two circuits. The flux generated by a circuit loop is proportional to the current flowing in the loop. Therefore, reducing the current can reduce coupling problems. As an example of the application of this idea, if a number of PCBs are connected via a backplane, the boards taking the greatest current could be positioned at the end of the backplane nearest to the power-supply feed, so that the loop of greatest area does not carry the full current.

Distributed-parameter circuits

So far, in discussing parasitic effects in electronic circuits, a conventional circuit approach has been used, supplemented in the case of capacitive and inductive coupling between circuits by some consideration of electric and magnetic fields. Circuit models are only an approximation to reality, and we might ask under what conditions a circuit model ceases to be valid, and what then happens? To answer these questions

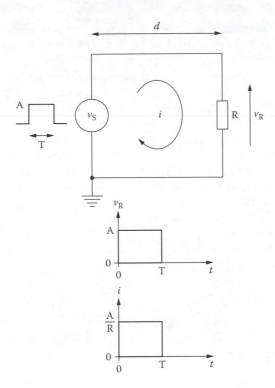

Figure 8.4   A pulse circuit to illustrate the importance of circuit dimensions.

it is helpful to look at the behaviour of transmission lines, because it is here that our conventional circuit models begin to fail us. The example presented below can easily be demonstrated in the laboratory and gives a good intuitive feel for the subject.

Figure 8.4 shows a circuit in which a voltage source generates a short pulse of amplitude, A, and duration, T. A load resistance, R, is connected to the source. Circuit theory tells us that the voltage, v_R, across the resistor will be identical to the voltage, v_S, at the source and that the current, i, will be a pulse of amplitude, A/R, and of the same duration, T, as the voltage pulse. In carrying out this simple analysis we have neglected, among other things, the parameter, d, the distance from the voltage source, v_S, to the load resistance, R, shown in the figure. To see why d can be important, imagine T to be 10 ns and d to be 10 m. How fast does the pulse travel along the wires from the voltage source to the load resistor? Well, it cannot travel faster than light, so we will take this to be the speed of the pulse. The speed of light is, to a good approximation, 3×10^8 m s^{-1} or, in more suitable units for electronic engineering, 300 mm ns^{-1}. The leading edge of the 10 ns pulse will therefore travel 3 m before the trailing edge leaves the voltage source, and the leading edge of the pulse will not reach R until 33 ns after leaving the voltage source. It follows therefore that R can have no effect on the current flowing from the voltage source, and our conventional circuit model fails. The reason is that the duration of the signal, T, is less than the size of the circuit, d, divided by the speed of light, c; that is,

In fact, the speed of an actual pulse can be as low as 100 mm ns^{-1}, or $\frac{1}{3}$ of the speed of light.

$$T < \frac{d}{c}. \tag{8.2}$$

Under these conditions, lumped-parameter circuit models in which circuit properties are considered to be localized in components are not valid, and we can no longer neglect the interconnecting wiring. Instead, we must consider the wiring as a transmission line and design the wiring to have the necessary transmission-line parameters for our application. The electrical energy travelling along a line can be thought of as an electromagnetic wave guided by the conductors.

Chapter 7 of Carter (1992) contains a much more thorough treatment of transmission lines than is presented here.

A transmission line has a characteristic impedance, Z_0, which depends on the geometry and dimensions of the line conductors and the permittivity of the medium or dielectric between the conductors. Z_0 is a dynamic impedance (it cannot be measured with a multimeter) and is independent of the length of the line. The input impedance of a transmission line is Z_0, and the output impedance or Thévenin equivalent source impedance is also Z_0. In the example shown by Figure 8.4, the current flowing from the voltage source is determined by the characteristic impedance of the transmission line linking the source to the load resistance and not by the value of the load resistance, R. As we shall see, the behaviour of the pulse on arrival at the load depends on the value of R and the value of Z_0, and some interesting effects can occur.

The relationship between Z_0 and the distributed parameters of the transmission line is given by

Kraus (1992) gave a derivation of Equation 8.3.

$$Z_0 = \sqrt{\frac{R + j\omega L}{G + j\omega C}} \tag{8.3}$$

where R is the series resistance per unit length (Ω m^{-1}), L is the series inductance (H m^{-1}), G is the shunt conductance (Ω^{-1} m^{-1}), and C is the shunt capacitance (F m^{-1}). On an ideal or lossless line, the series resistance and shunt conductance of the line are zero and Equation 8.3 reduces to $Z_0 = \sqrt{L/C}$. A practical transmission line has some resistance and conductance that absorb energy from any wave or pulse propagating on the line. The loss is quantified by an attenuation coefficient, α, which is typically 1 or 2 dB per 100 m, increasing with frequency.

The speed at which a wave or pulse travels down the line is determined by the relative permittivity of the medium or dielectric between the conductors. The propagation velocity is given by

$$v = \frac{1}{\sqrt{LC}} = \frac{1}{\sqrt{\varepsilon_0 \varepsilon_r \mu_0}} = \frac{c}{\sqrt{\varepsilon_r}} \tag{8.4}$$

where L and C are the distributed inductance and capacitance per unit length, ε_0 is the permittivity of free space, ε_r is the relative permittivity of the dielectric, μ_0 is the permeability of free space, and c is the speed of light.

Exercise 8.1

An RG58C/U coaxial cable has a characteristic impedance of 50 Ω and a shunt capacitance of 100 pF m^{-1}. What is (a) the series inductance, (b) the propagation velocity, and (c) the relative permittivity of the dielectric?

(*Answers*: (a) 250 nH m^{-1}, (b) 200 mm ns^{-1}, and (c) 2.25.)

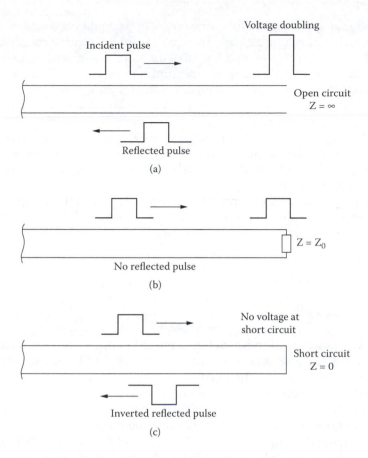

Figure 8.5 Transmission line termination effects: (a) voltage doubling at an open circuit, (b) matched termination, and (c) pulse inversion at a short circuit.

Exercise 8.2

What is the approximate propagation velocity on a PCB track if $\varepsilon_r = 6$ for PCB laminate?

(*Answer*: 120 mm ns^{-1}. In practice, the electric field extends outside the dielectric and the true velocity is a little higher.)

If a length of transmission line with characteristic impedance Z_0 is terminated with an impedance Z, what happens to a pulse travelling along the line on arrival at the termination? The answer is not only interesting but also of great significance in the design of high-speed digital systems. Figure 8.5 shows three special cases. In (a), the end of the line is left open circuit, and since no energy can be absorbed by an open circuit, the incident pulse is reflected and travels back along the line to the source. During the time that the pulse is being reflected, the incident and reflected parts of the pulse overlap so that the voltage present at the end of the line is twice the pulse amplitude. This effect is known as voltage doubling. In (b), the line is terminated with a resistance equal to the characteristic impedance of the line. There is no reflection, and this is thus a desirable situation if the line is carrying a signal from one part

There is also the possibility that the energy in the pulse could radiate from the open end of the line as an electromagnetic wave. In practice, a purpose-designed antenna is needed if any significant energy is to be radiated. A perfectly matched antenna would have an impedance of Z_0 and would radiate all the energy into free space.

of a digital system to another, since reflections could cause errors. The absence of reflection can be explained by imagining the terminating resistance to be replaced by a length of the same line extending to infinity. The input impedance of this semiinfinite line is Z_0. Once a pulse has been launched down an infinite length of line, we shall never see it again. An equivalent argument is that there is no reflection from a joint between two identical lines and therefore there is no reflection from a joint between a line and a resistance equal to Z_0. In (c), the line is terminated with a short circuit. As in case (a), no energy can be absorbed at the termination, and since there can be no voltage at a short circuit, the reflected pulse is inverted.

In the general case of a transmission line of characteristic impedance Z_0 terminated with an impedance Z, the amplitudes of the incident and reflected pulses are given by the equation

$$\frac{V_\mathrm{r}}{V_\mathrm{i}} = \frac{Z - Z_0}{Z + Z_0}. \tag{8.5}$$

Exercise 8.3

This is a practical exercise. Set up an oscilloscope of at least 50 MHz bandwidth, a pulse generator, and a length of 50 Ω coaxial cable as shown in the margin. The pulse generator must have a 50 Ω output so that any reflected pulses are absorbed when they arrive back at the pulse generator. With 30 m of cable the pulses will need to be about 100 ns duration, and with a shorter cable the pulses will need to be of shorter duration. Experiment with a variety of terminations including open and short circuits, 25 Ω, 50 Ω, and 100 Ω, and verify the behaviour discussed above. (The input impedance of the oscilloscope can be considered infinite if of 1 MΩ or more. Do not use an oscilloscope with a 50 Ω input. A 25 Ω termination is easily made from two 50 Ω terminations and a Tee-piece.)

Figure 8.6 shows some results obtained by the author using 6.5 m of cable and a 30 ns pulse as described in Exercise 8.3. Four cases are shown. The top trace in each case shows the voltage measured at the pulse generator, and the bottom trace shows the voltage measured at the far end of the cable. The pulse generator is not perfectly matched to the cable so that a secondary reflection occurs when the pulse arrives back at the pulse generator.

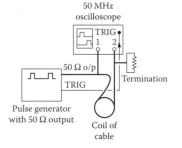

50 MHz oscilloscope

TRIG
1 2

50 Ω o/p

TRIG

Termination

Pulse generator with 50 Ω output

Coil of cable

Pulse propagation on transmission lines is very important in the design of high-speed digital systems.

Worked Example 8.3

Consider a parallel data link between two computers 5 m apart, made from ribbon cable with a characteristic impedance of 105 Ω and a capacitance of 50 pF m^{-1} between adjacent conductors. If handshaked data transmission is used, what is the maximum possible data transfer rate? (*Handshaking* means that the receiving computer acknowledges each data word by sending a signal back to the transmitting computer. Only when the acknowledgment is received does the transmitting computer send the next data word.)

Solution The capacitance is 100 pF m^{-1} if alternate conductors are grounded. Assuming the cable to be lossless and using $Z_0 = \sqrt{L/C}$ and Equation 8.4, the speed of propagation, v, is $1/Z_0 C$, which is $1/(105.100 \times 10^{-12})$ or 95 mm ns^{-1}. The propagation time over 5 m is thus 5000/95 or 52 ns, and, allowing for the acknowledgement, each data word will require about 100 ns, making no allowance for circuit propagation delays within the computers. This gives a data transfer rate of 10 M words per second.

Electromagnetic interference

Any unwanted electrical or electromagnetic energy that disturbs the normal operation of an electronic system can be classed as electromagnetic interference or EMI. Sources of EMI in industry include: electric motors, electric furnaces, arc-welding equipment, and radio-frequency heating and drying systems. In the home and the office, EMI can be caused by fluorescent lamps, dimmer switches, computers and wireless networks, thermostats, hairdryers, vacuum cleaners, and power tools. Military systems may also be exposed to radar transmissions and deliberate interference, or jamming, generated by hostile forces.

Some sources of EMI can be suppressed to reduce the level of interference. In many cases, however, equipment designers or manufacturers have no control over the environment in which their products will be required to operate, and the design must be capable of working in the presence of interference. Badly designed products can also be sources of interference, releasing unintended electromagnetic energy into the surrounding environment and causing an EMI problem elsewhere.

The study of EMI problems involving the effects of one system upon another, and within a complex system, is known as electromagnetic compatibility or EMC, and was first developed as a distinct discipline by electronics engineers working on military systems. EMC problems can occur in military aircraft, ships, and fighting vehicles because of the close proximity of many separate, sophisticated electronic subsystems for communication, navigation, target detection, range-finding, and weapon control. The increasing use of electronics in civilian applications made EMC an important aspect of all electronic product design, and this was recognized by the Commission of the European Communities (EC), which issued Directive 89/336/EEC in May 1989, known as the EMC Directive. It covered electrical and electronic apparatus defined in the broadest sense and required all such apparatus manufactured or sold in the then European Community from January 1, 1992, to comply with applicable European standards. Subsequently, because of delays in the completion of suitable standards, an amendment was issued to allow manufacturers to comply with relevant national standards until the end of 1995. After that date, compliance with European standards was mandatory and equipment satisfying these standards carries a mark consisting of the two letters *CE* (for Conformité Européenne), as will be explained in Chapter 11. Certification of compliance required testing, as will be discussed in Chapter 10. From July 2007 the directive is replaced by Directive 2004/108/EC, which reduces the requirement for

In the USA, the Federal Communications Commission (FCC) has issued rules governing electromagnetic emissions and Declaration of Conformity testing.

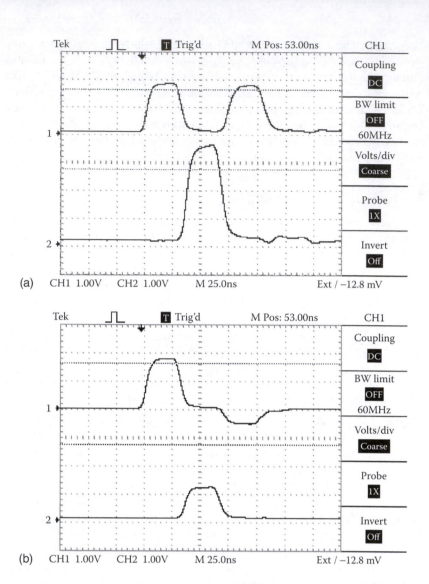

Figure 8.6   Pulse reflections on 6.5 m of 50 Ω coaxial line. Pulse width 30 ns. Top trace: transmitting end; bottom trace: termination end. (a) Open circuit termination, (b) 25 Ω termination, (c) 50 Ω termination, and (d) short-circuit termination. (Continued)

independent testing (because this proved to be burdensome) but imposes stricter requirements on documentation of the tests carried out.

EMI mechanisms

There are two main ways in which EMI can enter (or leave) an electronic system: conduction along wiring and cabling, and radiation through free space. Radiated EMI can be further divided into induction fields from nearby sources and plane-wave fields from far sources. Transmission line effects were introduced in the previous section, but so far the generation and reception of radiated EMI have not been discussed. Devices intended to launch or receive electromagnetic waves into or from free space are known as antennae. Circuits can exhibit parasitic antenna

Antennae are discussed by Kraus (1992).

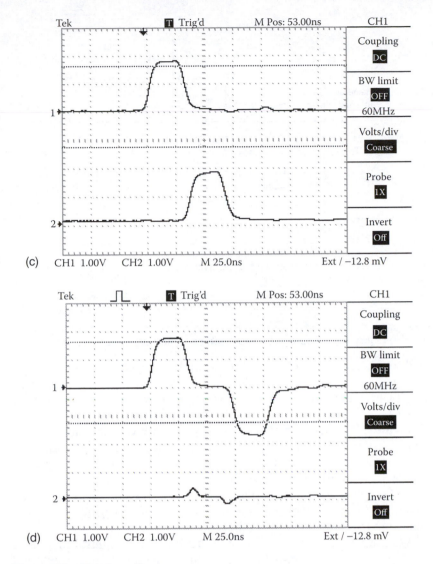

Coupling
DC

BW limit
OFF
60MHz

Volts/div
Coarse

Probe
1X

Invert
Off

(c)   CH1 1.00V   CH2 1.00V   M 25.0ns   Ext / −12.8 mV

Tek   T Trig'd   M Pos: 53.00ns   CH1

Coupling
DC

BW limit
OFF
60MHz

Volts/div
Coarse

Probe
1X

Invert
Off

(d)   CH1 1.00V   CH2 1.00V   M 25.0ns   Ext / −12.8 mV

Figure 8.6   (Continued)

effects and may generate not only local electric and magnetic induction fields, which can couple into other nearby circuits, but also electromagnetic waves, which can propagate farther away and cause interference to distant systems. Parasitic antennae may also receive interference from distant sources.

Far from an antenna, in terms of signal wavelength, the electromagnetic waves generated have a constant ratio of electric field strength to magnetic field strength of 377 Ω. This is known as the intrinsic impedance of free space. From electromagnetic theory, it can be shown to be given by $\sqrt{\mu_0/\varepsilon_0}$, where μ_0 and ε_0 are the permeability and permittivity of free space respectively. Close to an antenna, in the near field, the ratio of electric field strength to magnetic field strength or the field impedance can be less than or greater than 377 Ω depending on the type of antenna, or circuit, generating the field. High-impedance fields, where the electric field component dominates, are generated by

These statements should be compared with those made earlier in the discussion of parasitic capacitance and mutual inductance. They are simply another way of looking at the same phenomenon.

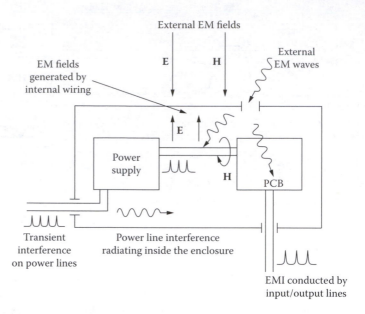

Figure 8.7   Routes by which EMI can enter or leave a system.

high-impedance circuits operating at high voltages and low currents. Low-impedance fields with a strong magnetic component are generated by high-current circuits with low series impedance and low voltage drops.

Figure 8.7 depicts the main routes by which EMI can enter an electronic system. All of these routes are reversible, and EMI generated by the system can leave along the same pathways to cause trouble elsewhere. It is important to realize that interference conducting along, say, mains supply wiring may radiate once inside the equipment and that radiated EMI may couple into internal wiring and propagate by conduction from one subsystem to another.

We are now in a position to study techniques to control or reduce EMI problems, first from the viewpoint of susceptibility to external EMI and second from the EMC viewpoint in reducing EMI emissions from a system. As we shall see, there are techniques that are helpful from both points of view.

Reduction of EMI susceptibility

When we are designing an electronic product or system, we may not have a detailed knowledge about the EMI environment in which our design is to work. This can make reduction of EMI susceptibility (or improvement of EMI immunity) a somewhat empirical activity. In some designs, there will be specific EMI sources nearby operating on known frequencies and at known power levels. Protection measures against interference can then be designed with greater confidence using EMC manuals and calculation to analyse the performance of a proposed protective measure. There is not sufficient space in this book to discuss such matters, and the material that follows is intended only to outline the general principles.

A more thorough treatment is given by Chatterton and Houlden (1992), who discussed EMC in detail and gave references to specialist literature. Williams (2001) presented a more pragmatic approach and discussed relevant EMC standards and legislation.

136

Before discussing techniques for preventing EMI entering a system, it is worth examining the possibility of designing a system to be inherently immune to EMI. One way in which this can be done is to limit the bandwidth of the system to just that range of frequencies necessary for the system to do its job. A thermocouple amplifier, for example, can be limited to a response time of the same order as that of a thermocouple, say 1 s. The frequency response of the amplifier can cut off therefore at less than 10 Hz, so that sensitivity to 50/60 Hz and r.f. interference will be reduced. Another possibility for reduction of mains frequency interference can be applied in the design of an averaging measurement system. If the averaging time is designed to be a multiple of the mains supply period, any interference at the mains frequency or its harmonics will average to zero over the measurement time. This technique can be used with a dual-slope integrator analogue-to-digital converter (ADC), as shown previously in Figure 6.3. This type of ADC integrates its analogue input signal over a time, T_1, determined by a digital counter and a clock. The integrated voltage is then effectively compared with a reference voltage to produce the digital result. Since T_1 is dependent only on the counter length and the clock frequency, a value that is an integral multiple of the mains period can be chosen. Any mains hum present in the analogue input signal is thus averaged out during the first stage of the conversion so that the integrated input voltage, and thus the final digital result, is independent of mains hum.

Any mains hum present in the dual-slope ADC itself is not of course averaged out.

Grounding

In circuit design, we normally take for granted the concept of an earth, ground, or 0 V reference potential. This concept is worth examining in a little detail, as a practical ground will depart from the ideal. Consider a signal source connected to a load as shown in Figure 8.8. The source might represent a transducer such as a microphone producing a small voltage, say 10 mV, while the load might represent the input impedance of an amplifier. There are three potential differences shown in the figure, of which v_G would be zero in a circuit with an ideal ground, leaving only the transducer electromotive force (e.m.f.) and the voltage at the amplifier input, which would be equal by application of Kirchoff's voltage law. In reality, of course, the ground and signal connections have some nonzero impedance, and a potential difference can exist along their

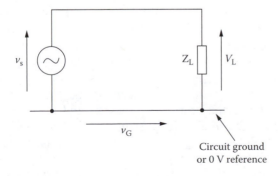

Figure 8.8   A signal source connected to a load.

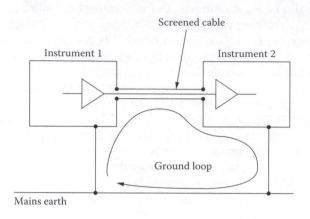

Figure 8.9   The ground loop problem.

lengths if a current flows. Ground currents can be caused by external electromagnetic induction effects and by power-supply return currents. Both can be controlled by careful layout and design of a grounding system, or by avoiding grounding altogether and using a floating ground, which is simply a 0 V reference that is electrically isolated from mains or other earths.

Single-point grounding is an ideal arrangement, in which all ground connections within a system are made at a single point, so that from any circuit in the system there is only one path to earth. This avoids the creation of ground loops in which a closed loop of ground conductor exists. A ground loop can be coupled by induction fields, especially magnetic fields from nearby mains transformers, causing circulating currents and spurious voltages in the ground loop. Figure 8.9 shows a typical situation in which a ground loop exists and where the ideal of single-point grounding cannot be attained. Two electronic instruments are connected to mains earth by their mains supply leads, and the chassis or cabinet of each instrument is earthed for safety reasons as well as to shield against EMI. A signal is passed from one instrument to the other via a screened cable, and the cable screen is connected to the cabinets of both instruments, forming a ground loop as shown. Circulating currents in the ground loop may cause a potential difference along the length of the cable that adds to the signal voltage measured by instrument 2. There are various solutions to this problem depending on the frequency range and sensitivity of the instruments. One possibility is to separate the signal grounds within the instruments and the screen of the cable from mains earth with a resistance so that the impedance of the signal ground loop is increased. The obvious possibility of breaking the ground loop at some point cannot normally be used: both instrument cabinets must be earthed for safety reasons and if the cable screen is disconnected at one end, the signal return current must follow the mains earth path, which has a large self-inductance because of the loop area. A second possibility is to provide a high-quality, low-impedance ground connection between the two instruments using a heavy copper braid strap.

138

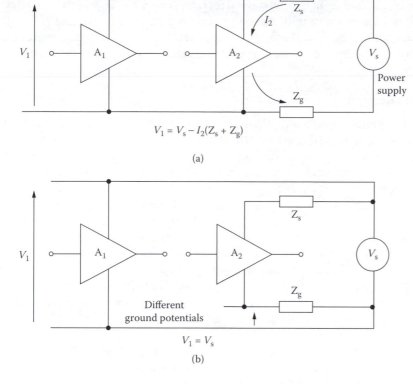

$$V_1 = V_s - I_2(Z_s + Z_g)$$

(a)

$$V_1 = V_s$$

(b)

Figure 8.10   Power-supply coupling.

At higher frequencies, multipoint grounding must be used, because when the length of a ground lead approaches a quarter of the signal wavelength, the lead no longer provides a low-impedance path to ground. Ground leads must therefore be kept short and connected directly to a ground plane at the nearest available point. The multipoint ground must be of low impedance and carefully designed to maintain a low impedance over the operating life of the system. Materials must be chosen, for example, so that corrosion does not impair grounding effectiveness.

Power-supply coupling

A signal in one circuit can constitute interference if it gets into a neighbouring circuit. Intercircuit coupling can occur via the power supply circuit where two circuits share a power supply. Figure 8.10 shows how power-supply coupling can occur. In (a) two amplifiers, A1 and A2, share a common power supply such that A1 is downstream of A2. Z_s and Z_g are the impedances of the power supply and ground return conductors between the power supply and A2. I_2 is the current drawn by A2 and will have a steady component due to internal bias currents and a fluctuating component due to the signal being amplified by A2. As shown, the supply voltage V_1 at A1 is reduced by the voltage drop across Z_s and Z_g, so that a signal component from A2 can enter A1 via the power supply leads. Figure 8.10b shows a power distribution arrangement designed to overcome this problem. A2 is supplied by separate leads running back to the power supply, or at least to a point where a low-impedance path exists back to the power supply. The voltage drops across Z_s and Z_g now no longer affect the supply to A1, but there is now a possible problem with differing ground potentials at A1 and A2, and we could be back to the problem of providing a low-impedance ground return to reduce Z_g. Practical engineering problems are very often like this!

Filtering

Interference conducted along mains and other cables and internal wiring can be reflected or attenuated by filters, which allow wanted frequencies to pass while blocking unwanted frequencies. Reflective filters contain reactive components only and block conducted EMI by reflecting unwanted energy back along the wiring, in a similar manner to unmatched transmission-line terminations. Lossy filters, on the other hand, also have resistive elements (not necessarily in the form of lumped resistors) to absorb unwanted energy and dissipate it as heat.

Conducted interference is likely at frequencies of up to 30 MHz. Above this frequency, interference tends to radiate from conductors rather than propagating as a guided wave. High-frequency interference does not radiate significantly from a purpose-designed high-frequency cable such as a coaxial cable, so that interference above 30 MHz can be conducted well by such a cable.

Mains filters with integral mains connectors of the type shown in Figure 8.11 can be used to filter mains-borne interference. The construction of this type of filter means that there is no unfiltered mains wiring inside the equipment enclosure, eliminating radiated interference problems from the mains supply wiring. A common circuit for a single-phase mains filter is shown in Figure 8.12. The operation of this circuit can be considered as a deliberate transmission line mismatch, or in terms of the conventional circuit mode model. The inductors, or chokes, L_1 and L_2, are wound on a common toroid so that the magnetic fluxes generated by normal a.c. currents passing through the filter cancel out and ensure that the core does not saturate. This means that the inductors present an unavoidable low impedance to differential-mode interference where the line and neutral conductors carry equal and opposite interference currents. Common-mode interference, however, is blocked by the high impedance of the two chokes to currents flowing equally in the

Figure 8.11   A commercial mains inlet filter with integral mains inlet connector to IEC 60320 C16. (Courtesy of Schaffner EMC Ltd.)

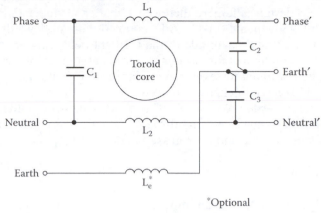

*Optional

Figure 8.12   Typical equivalent circuit of a mains filter.

Capacitors connected between phase/neutral and earth, as shown in Figure 8.12, may be damaged by high-voltage insulation resistance tests, as discussed in Chapter 11.

two supply lines. The filter shown in Figure 8.11, for example, gives an attenuation or insertion loss of only 10 dB at 150 kHz to a difference signal, but over 40 dB at the same frequency to a common-mode signal, when tested using a 50 Ω source and a 50 Ω load. The capacitors shown in Figure 8.12 present a low impedance to interference currents and further reduce the proportion of energy passing through the filter. The optional earth-line choke, L_e, blocks high-frequency interference travelling along the earth line.

At high frequencies, the earth line cannot be considered as a negligible impedance to earth. High-frequency interference can therefore propagate along the earth conductor.

The performance of a filter can, in principle, be stated by giving the attenuation or insertion loss as a function of frequency. There is a difficulty, however, in that the performance of the filter depends on the impedances of the source and load networks, and these vary from one application to another. Tests may be needed, therefore, to establish the performance of a particular filter in a given application.

Data and signal cables can be filtered at the point where they enter a system using filtered connectors with integral distributed filter elements. These can take the form of a capacitive sleeve connected to ground around each pin of the connector, or a lossy ceramic or ferrite tube with conductive coatings on the inner and outer cylindrical surface, behaving as a lossy transmission line.

Ferrite or lossy ceramic tube

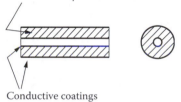

Conductive coatings

A distributed filter element.

For r.f. applications, leadthrough or feedthrough capacitors are used to pass d.c. power through a bulkhead without allowing r.f. energy to pass.

Shielding

If we wish to exclude electromagnetic fields from a region inside an electronic system, we can surround the region with an electromagnetic shield. Static electric fields are easily excluded from an equipment enclosure by a conducting screen or Faraday cage, but, unfortunately, radiated electromagnetic fields are time varying and have both electric and magnetic components. A static electric field is always perpendicular to the surface of a conductor, cannot penetrate the conductor, and cannot cause current to flow in the conductor. Time-varying electromagnetic fields, however, can penetrate a conducting shield and can induce currents in the shield. The degree of penetration varies with the frequency of the field and the type of shielding material.

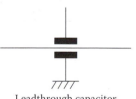

Leadthrough capacitor

141

Shielding design is largely therefore a matter of understanding the mechanisms by which a shield works coupled with a knowledge of the type, frequency, and magnitude of the incident field. Tabulated shielding effectiveness figures for various materials at a range of frequencies can then be consulted to select a material or combination of materials and to decide on the thickness required.

The shielding effectiveness, S, of a shield is defined as the ratio of incident power to power passing through the shield. S is made up from three components that add if expressed in decibels:

$$S = R + A + B \tag{8.6}$$

R is a reflection loss due to an impedance mismatch between the shield and the incident field, and depends on the shield material, the field frequency, and the field impedance. A is an absorption loss and depends on the shield material and field frequency, but not the field impedance. B is a correction term that accounts for reflection from the inner surface of the shield. R is significant for high-impedance and plane-wave fields so that a thin layer of, say, copper can provide effective shielding. R for copper at 10 MHz is about 100 dB for plane-wave fields. For low-impedance fields, there is less of a mismatch between the shield and field impedance, and S must be made up mainly from the absorption loss A.

Practical shields have weaknesses such as joints and openings that complicate estimates of shielding effectiveness. Slots in a shield, for example, can act as antennae and reradiate energy from circulating currents inside the shield into the volume enclosed by the shield. Removable covers and lids create narrow slots where r.f. energy can pass through the shield, and they may need to be sealed with radio-frequency gaskets of knitted wire mesh or conductive foam polymers.

Suppression

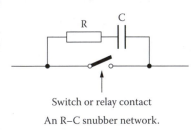

Switch or relay contact

An R–C snubber network.

Motors and contact devices such as switches and relays can generate EMI in the form of transient or short-duration voltages, currents, and electromagnetic fields because of electrical arcing. Many types of switching circuits also generate transient interference. Empirical techniques exist to control or suppress this type of EMI including resistor and capacitor (R–C) snubber networks across switch and relay contacts and suppression capacitor assemblies for suppression of motors.

Measures taken to reduce the high-frequency content of signals, such as limiting the rise time of pulse edges, can also be regarded as suppression.

Applications studies

This section presents studies of several applications where parasitic electrical and electromagnetic effects have an important influence on performance and design, both to illustrate the types of problems that can be met with and to show how good engineering practice can be justified in terms of the effects outlined so far in this chapter.

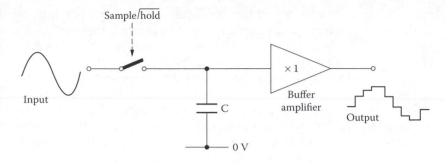

Figure 8.13   A sample-and-hold (S–H) circuit.

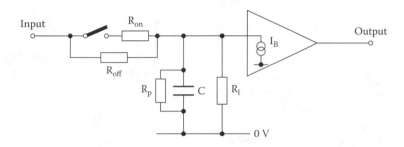

Figure 8.14   The circuit of Figure 8.13 with some of the parasitic circuit elements added.

A sample-and-hold circuit

Figure 8.13 shows a common circuit in which parasitic effects are an important influence on performance. A sample-and-hold (S–H) circuit samples or captures the voltage at its input and holds that voltage at its output until the next sample is taken. They commonly precede ADCs in systems where an analogue signal, such as speech, is to be converted to digital form for long-distance transmission or signal processing. The operation of this circuit is, in principle, very simple. To sample the input signal, the switch, which is in practice a field-effect transistor (FET), is closed. The capacitor, C, charges up to the value of the input voltage and will follow the input voltage as long as the switch remains closed. To hold the sampled value, the switch is opened, isolating the capacitor C from the input and allowing the capacitor to hold the sampled voltage. A unity-gain buffer amplifier provides output current to supply a load, such as the input of an ADC, without drawing current from the capacitor. In an ideal S–H circuit, the capacitor would charge instantaneously when the switch was closed and would hold the stored voltage precisely for an indefinite time when the switch was open. In a real circuit the time taken to charge the capacitor, or the acquisition time, is not zero and the charge stored on the capacitor leaks away when the switch is open. The output voltage is said to droop during the hold state. The droop rate, in mV s^{-1}, is an important performance parameter in high-resolution analogue-to-digital conversion applications. How can we estimate these parameters, and what design steps could we take to improve the performance of this circuit?

Figure 8.14 shows the same circuit, but with some of the parasitic properties added. The switch cannot be perfect and will have a small

Dielectric absorption, discussed in Chapter 5, is also a significant parasitic component effect in an S–H circuit, and also influences the choice of capacitor type.

but nonzero resistance, R_{on}, when closed and a large but not infinite resistance, R_{off}, when open. The capacitor has a parasitic dielectric leakage resistance, R_p, which can be reduced by choosing a suitable type of capacitor, and the buffer amplifier has an input bias current, I_B. All of these parasitic effects are properties of the components and can be reduced by careful choice of components. R_l, however, is a leakage resistance to 0 V due to the printed circuit board material, which can be increased by careful layout design to make leakage paths as long as possible and perhaps by using a solder resist to reduce moisture absorption by the PCB laminate, which would otherwise reduce R_l.

Worked Example 8.4

A precision S–H integrated circuit (IC) with 1 nF hold capacitor has a stated maximum droop rate of 30 mV s⁻¹. If the maximum input voltage is 5 V, what is the total leakage current from the capacitor? The hold capacitor is connected to an external pin of the IC. What order of insulation resistance is needed on the PCB if the droop rate is not to be significantly increased?

We are assuming a constant leakage current, a valid approximation since the hold voltage is normally digitized by a following ADC within a few milliseconds.

Solution The capacitor voltage and the leakage current are related by the equation $i = C \, dv/dt$.

Since dv/dt is 0.03 V s⁻¹, the leakage current must be 30 pA. At a maximum hold voltage of 5 V, this corresponds to a leakage resistance of 5 V/30 pA, which is 167 GΩ (1.67×10^{11} Ω).

If the droop rate is not to be significantly increased by PCB leakage, the insulation resistance between the hold capacitor and ground on the PCB must be, say, five times higher or about 800 GΩ. This is a very high resistance value, and it is unlikely to be achievable.

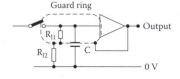

The leakage current through R_l can be reduced by adding a guard ring around the capacitor terminal as sketched in the margin. The guard ring is held at the same potential as the capacitor by connecting it to the buffer amplifier output. The leakage resistance, R_l, is split into two series resistances, R_{l1} and R_{l2}, as shown, and since the potential difference across R_{l1} is zero to a close approximation, the leakage current that flows from the capacitor is negligible. This technique therefore solves the problem of the very high value of leakage resistance that would otherwise be needed.

Exercise 8.4

(Discussion point.)

If a guard ring is added to the S–H circuit in Figure 8.13 by forming a ring from the PCB track around the capacitor terminal, would it be a good idea to use a solder-resist coating on the PCB?

Analogue systems

Despite the modern trend towards digital techniques in many applications areas of electronics, analogue circuits are still indispensable for conditioning and amplifying low-level signals from transducers, antennae, and detectors. New active devices and IC techniques tend to encourage the development of systems of greater sensitivity. Problems with EMI susceptibility can occur therefore in many analogue systems. EMC emission problems are less likely unless a system operates at radio frequencies, or with high voltages or currents. R.f. and microwave circuit construction requires special techniques that have been excluded from this book.

Good engineering practice in analogue system design is to limit the operating bandwidth of the system, as already described. We should be aware, however, that measures taken to limit the bandwidth may not be effective at higher frequencies as the parasitic properties of components depend on frequency. High-frequency roll-off in an amplifier, for example, depends on strong negative feedback, which may depend on a feedback capacitor. If the series inductance of the capacitor becomes significant at higher frequencies, the feedback provided by the capacitor may cease to hold down the amplifier gain.

Even when the bandwidth of a system has been limited to the minimum necessary, EMI falling outside the system bandwidth may be translated into an in-band signal by nonlinearities in the system in the same way that an amplitude-modulated r.f. carrier can be demodulated by a p–n junction. Two frequencies can also be combined by intermodulation, to produce sum and difference frequencies, by nonlinear components acting as spurious analogue multipliers. In-band interference generated by frequency translation can be a serious problem.

Exercise 8.5

Given that two voltage sources, $V_1 \sin \omega_1 t$ and $V_2 \sin \omega_2 t$, are connected in series to a nonlinear device with a characteristic given by $I = kV^2$, where k is a constant, show that the device current will contain sinusoidal components at frequencies of $(\omega_1 + \omega_2)$ and $(\omega_1 - \omega_2)$.

Analogue circuits are often built on ground-plane PCBs. Double-sided PCB construction is the simplest and cheapest method of building a ground-plane board. The component side of the board is covered with copper except where component leads pass through the board. This side of the board is the ground plane and serves as the 0 V reference for the whole circuit. Most, or all, of the tracks are laid out on the underside of the board. Figure 8.15 shows this type of construction for classic through-hole mounted components, and Table 8.1 lists the main advantages and disadvantages of a ground-plane board.

Table 8.1   Advantages and disadvantages of ground-plane PCBs

(a) Advantages

1. Reduced parasitic capacitance between components and tracks.
2. Reduced EMI susceptibility.
3. Low-impedance ground return to power supply.

(b) Disadvantages

1. Increased parasitic capacitance to ground.
2. Availability of only one side of the board for tracks.

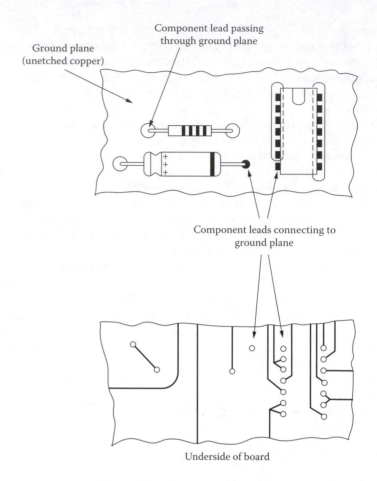

Figure 8.15 A ground-plane PCB.

Careful attention to filtering, using R–C networks and ferrite beads on power-supply rails and bias networks and on signal paths between stages of a circuit, is essential to control power-line interference and to reduce the amplitude of EMI at the earliest possible stage in the system, before frequency translation or amplification occurs. In an amplifier system, signal feedback from later stages to earlier stages must also be controlled, including feedback through power-supply rails by power-supply coupling as described earlier in this chapter.

Digital systems

Digital electronic circuits are very widely used in computers, control systems, communications, instrumentation, and signal processing. They often include microprocessors and other digital circuits operating at clock frequencies of hundreds of megahertz and with pulse risetimes below 1 ns. A moderately sized PCB of double Eurocard format (233.4 × 160 mm) can accommodate 100 or more ICs, sufficient to draw a power-supply current of several amperes. Digital PCBs and systems can therefore be troublesome sources of EMI, generating low-impedance, predominantly magnetic, local fields and far-field emissions at frequencies of over 100 MHz. EMI susceptibility is not often a problem with digital systems because they are inherently immune to spurious voltages of several hundred millivolts or more. Conducted EMI entering a system is most likely to be dealt with adequately by measures such as filtering taken to control conducted EMI leaving the system.

Mixed analogue and digital systems such as microprocessor-based instruments can present a serious EMC problem if the analogue parts of the system are processing low-level signals, as is frequently the case. Conducted and induced EMI from the digital circuits must be contained and controlled. Separate power supplies and grounds for the analogue and digital parts of the system are advisable, and care should be taken to segregate the analogue and digital circuits.

High-speed digital systems operating at clock frequencies of hundreds of megahertz and with pulse risetimes below 1 ns are excellent examples of distributed-parameter systems where the finite speed of light is important. Consider, for example, a single logic gate on a PCB. We have seen that the propagation velocity of electromagnetic signals on a PCB transmission line is about 150 mm ns^{-1}. Time delays in signal propagation are not likely to be a major problem unless our system is operating at a clock frequency of hundreds of megahertz, provided the system is reasonably compact. Time delays in energy propagation from the power supply to a logic gate can cause problems, however, if the gate switches rapidly compared to the propagation time from the power supply to the gate. Figure 8.16 illustrates the problem that can occur, especially with classic transistor–transistor family logic. (The problem described here is worst with TTL, but present even in systems based on CMOS logic.) For simplicity, a basic TTL circuit is shown, but the same considerations apply with more commonly used types of TTL logic such as advanced low-power Schottky (LS TTL). When the output of gate 1 is high, a small reverse current of 40 μA flows into the emitter of the input transistor of gate 2. When the output of gate 1 switches to low, however, a current of 1.6 mA flows out of the input of gate 2 as shown. If gate 1 is driving the maximum allowable number of gates (10), gate 1 sinks 16 mA as shown. This current must eventually be supplied by the power supply, but initially the current must be supplied locally to gate 2 until

The sketch below shows the gate-output and gate-input logic levels for classic 4000 series complementary metal-oxide semiconductor (CMOS) logic operating at its normal maximum supply voltage of 12V. A worst-case logic low level at an output is 50 mV, which is 3.95 V less than the worst-case logic low threshold at a gate input. A spurious voltage of almost 4 V must therefore be added to a gate output in the low state to cause an error. This value is called a noise margin. (Most modern systems operate at 5 V or less, in which case the noise margins are lower.)

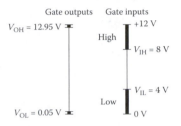

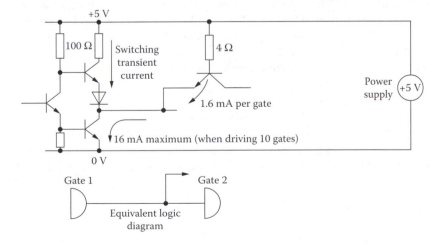

Figure 8.16   The switching transient problem in TTL family logic circuits.

a signal travels from gate 2 to the power supply and back. If there is no local source of current, the rail voltage in the vicinity of gate 2 will fall, possibly causing spurious logic transitions in nearby gates. The problem is made worse because both output transistors of gate 1 conduct during the output transition, drawing a transient current from the supply rail. If a TTL circuit is to operate reliably, therefore, local reservoirs of charge are needed close to each logic gate in the form of decoupling capacitors. Traditionally, 100 nF ceramic capacitors were used, but there is no reason why other types should not be suitable provided charge can be extracted sufficiently rapidly from the capacitor.

The term *decoupling* is historic and does not accurately describe the function of these capacitors. They are charge reservoirs.

After local capacitors have supplied the transient current demand caused by a logic transition, the power supply must take over to supply the steady current and replenish the local reservoirs. The self-inductance of the power-supply loop now becomes important: a large open loop of power rail and return wiring will have a larger self-inductance than a small loop where supply and return conductors are both short and close together throughout their length, and the self-inductance of the large loop will restrict the increase in current and cause a voltage drop along the supply and return conductors. Depending on the distance to the power supply, there may be a need for a few larger reservoir capacitors on a PCB to supply current after the local decoupling capacitors have supplied the immediate transient demand but before the power supply has had time to respond.

The opposite situation to that described above occurs when there is a drop in current demand due to a logic transition: the reservoir capacitors have to absorb surplus current until the power supply responds and reduces its output current.

Similar problems (sudden variation in current demand) can occur with microprocessors and other digital circuits that dynamically vary the clock frequency applied to internal subsystems, or that cut off the power supply to subsystems that are temporarily idle (to save power). For example, a processor with a floating-point unit may apply power to this unit only when a floating-point operation is being performed.

We are now in a position to look at examples of good engineering practice in digital PCB design. The first example is multilayer construction with a minimum of two internal layers for supply rail and ground planes, and with signal conductors routed on the outer faces of the board. Table 8.2 lists the main advantages and disadvantages of multilayer construction. Decoupling capacitors are still required (depending on the logic technology being used), even though the PCB itself contributes some distributed capacitance.

Exercise 8.6

Estimate the total capacitance between power and ground planes of a double Eurocard (233.4 × 160 mm) multilayer PCB with a plane spacing of 0.3 mm and $\varepsilon_r = 6$. $\varepsilon_0 \cong 9$ pF m^{-1}. Comment on the magnitude of the result.

(*Answer*: about 7 nF.)

For comparison, we consider the advantages and disadvantages of the cheaper double-sided construction for a digital board. This approach is

Table 8.2   Advantages and disadvantages of multilayer construction for digital PCBs

(a) Advantages

1. Low-impedance power distribution.
2. More uniform Z_0 for signal tracks.
3. Reduced EMI susceptibility because of low circuit loop area (all conductors are close to the ground plane).
4. Reduced EMI emission because of slower edges on logic transitions, caused by distributed capacitance, and low loop area.
5. Reduced crosstalk among tracks because of proximity to ground plane.

(b) Disadvantages

6. Higher manufacturing costs compared to double-sided plated-through hole (PTH) boards.
7. Difficult to rework and repair.

Table 8.3   Advantages and disadvantages of double-sided construction for digital PCBs

(a) Advantages

1. Low manufacturing cost.
2. All connections accessible for rework and repair.

(b) Disadvantages

3. Careful attention to layout needed, especially in power distribution.
4. Susceptible to radiated EMI because of open structure of circuit loops.
5. Significant EMI emissions from fast edges and clocks.
6. Crosstalk between adjacent closely spaced tracks.

less used nowadays than it was in the 1980s because clock speeds are now much higher, and the costs of multilayer boards have fallen as they have become more widely used. Table 8.3 lists the main advantages and disadvantages of double-sided construction, assuming that there is no significant area of ground plane on the board.

Summary

Many problems with electronic circuits and systems can be attributed to parasitic electrical and electromagnetic effects. The symptoms of these effects can include crosstalk, instability, pickup, and interference. The causes of these symptoms include parasitic properties in components, parasitic circuit elements, and electromagnetic coupling and radiation.

The finite speed of light, which is also the maximum speed at which electrical and electromagnetic signals and energy can propagate, influences the behaviour of an electronic system whose dimensions are greater than the speed of light multiplied by the signal period or time duration. Under these circumstances, lumped-parameter circuit models are no longer valid and conductors must be considered as transmission lines. The speed of signal propagation on a transmission line is $c/\sqrt{\varepsilon_r}$, where c is the speed of light and ε_r is the relative permittivity of the transmission-line dielectric. There is no signal reflection from the end of a transmission line terminated with its characteristic impedance.

Unwanted electromagnetic energy entering a system is known as electromagnetic interference or EMI and can propagate by conduction along wires and cables or by radiation through free space. Parasitic antennae in electronic systems can radiate or receive electromagnetic energy. Close to an antenna in the near-field region, the electromagnetic field may be predominantly electric or predominantly magnetic. Farther away in the far-field region, the energy in the field becomes divided equally between the electric and magnetic components and the field impedance equals the impedance of free space. The study of electromagnetic interactions between electronic systems and subsystems is known as electromagnetic compatibility or EMC.

Measures to reduce the EMI susceptibility of a system can include a limited bandwidth, well-designed grounding, filtering, and shielding. EMI emissions from a system can be reduced by suppression, grounding, filtering, and shielding.

Problems

8.1 Show that the output of a transmission line carrying a voltage signal, $V_i(t)$, can be modelled as a Thévenin equivalent network consisting of a voltage source, $2V_i(t)$, in series with an impedance, Z_0.

8.2 Two PCB tracks with a characteristic impedance of about 100 Ω run alongside each other for a distance of 100 mm. One track carries a 10 MHz square-wave clock signal. The parasitic capacitance between the tracks is 10 pF m^{-1}. The Fourier series for a square wave of amplitude $\pm V_p$ is

$$\frac{4}{\pi} V_p \sum_{n \text{ odd}} \frac{1}{n} \sin n\omega_0 t$$

where ω_0 is the fundamental frequency (here 10 MHz). (The square wave is composed of odd harmonics.)

a. What is the highest frequency at which the capacitive coupling between the tracks can be modelled as a lumped-parameter circuit? Assume a propagation velocity of 150 mm ns^{-1}.

b. By treating the parasitic capacitance as if it were lumped, and the tracks as 50 Ω source and load networks, use Equation 8.1 to find the interference voltage, expressed in decibels relative to the square-wave amplitude, coupled from the clock track to the other track at all harmonics up to the ninth (90 MHz). Is this a fair approximation?

8.3   A data bus for a 1970s digital computer was designed so that each conductor had a characteristic impedance of 120 Ω. Open-collector logic was used to drive the bus, and resistor networks were used at each end of the bus, as shown, to give a Thévenin source voltage of 3 V and a resistance of 120 Ω.

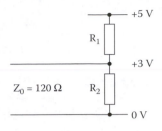

    a.   What values of resistor are required?

    b.   What is the maximum data rate achievable over a 3 m length of bus if the transmitter is at one end and the receiver is at the other end, and the receiver acknowledges each datum transferred? ($\varepsilon_r = 6$.)

8.4   An 8-bit bus driver IC drives eight 120 Ω transmission lines.

    a.   If all eight gates drive their outputs to 3 V simultaneously, what is the local instantaneous current demand?

    b.   If the power supply is 300 mm away and signals propagate at 200 mm ns^{-1}, what is the absolute minimum local capacitance required to prevent a fall in rail voltage of more than 50 mV?

    c.   Why, in practice, is the answer to (b) optimistic?

Reliability and maintainability

9

Objectives

- [ ] To introduce reliability and maintainability.
- [ ] To discuss the meaning of the term *failure*.
- [ ] To introduce the "bathtub" curve.
- [ ] To introduce quantitative measures of reliability and maintainability.
- [ ] To outline the principles of high-reliability systems.
- [ ] To discuss maintenance and design for maintainability.

All electronic equipment has a limited life. The materials and components from which products are fabricated gradually deteriorate through wear or through physical and chemical processes such as creep, fatigue, diffusion, corrosion, and embrittlement, leading eventually to failure. By replacing worn, aged, and faulty components, equipment life can be extended, but eventually this becomes more expensive than total replacement.

It is possible to estimate the probability of a product continuing to function without failure over a specified time period or amount of use (the reliability of the product) by using data from life tests on samples of the product or from data on the reliability of the product's components. It is not, however, possible to predict the actual time of failure of an individual specimen since there is a degree of randomness inherent in failure mechanisms. Reliability is therefore a statistically based subject.

Modern electronic components, especially integrated circuits, are highly reliable, particularly when their complexity is taken into account. This has encouraged the widespread application of electronic systems in industry, offices, and homes. In the past, many electronic products were designed to a functional or performance specification with little consideration of reliability or maintainability. This was possible because electronic systems of moderate complexity operating under benign conditions are inherently reliable. This, in turn, encouraged the development of more complex products and the introduction of electronics into more hostile environments where operating stresses are more severe. For example, road vehicles, where electronics have been applied to ignition and engine management systems, are a particularly arduous environment, providing vibration, extremes of temperature, and electrical interference that can cause rapid failure of a poorly designed system. Electronics is also widely applied in safety-critical applications such as life support, railway signalling, and aircraft control systems (fly-by-wire). In these and other applications, high reliability is important.

Improved product reliability and maintainability can be economically worthwhile. Reliable, easily maintained products have a lower cost of ownership partly because of the low maintenance costs but also

The word *reliability* is used both in the sense defined here and also as the name of a subject or discipline concerned with failure and the analysis and prediction of failure.

Two useful general references for this chapter are Cluley (1981) and O'Connor (2002). Many of the technical terms used in this chapter are defined in British Standard 4778.

Design for function was normal in the 1950s and early 1960s, when electronic products were relatively expensive compared to the labour costs involved in repair. Today, electronic products are cheap compared to labour costs, and the cost of repair is much more significant.

because they are available for use for a greater proportion of the time than an unreliable product that requires extensive maintenance. Manufacturers of a product can also benefit from improved reliability and maintainability because they can reduce their support resources for the product by stocking fewer spare parts and by having fewer staff and less equipment devoted to repairs on faulty units. This is especially so if the product is large and complex and needs to be repaired by field service staff visiting the customer's premises or site. Complex products such as railway vehicles are often monitored remotely over telephone lines using modems, or over Internet or private network links, to ensure that scarce and expensive field service staff visit the customer's site equipped with the correct replacement units when a fault develops.

Failure

In the calculator example, the failure of the divide function is a complete failure of that function, but not a total failure of the calculator. If a reciprocal key is available, continued use of the calculator is possible by using the reciprocal and multiplying functions to implement division.

Drift failure

Monotonic drift

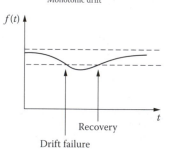

Recovery

Drift failure

Nonmonotonic drift.

Failure of an electronic system or product normally means that the system is no longer able to perform one or more of its normal functions. Failure of an oscilloscope, for example, might mean that no trace was displayed for one of the vertical channels. A fault in a pocket calculator could result in loss of the divide function but not the add, subtract, and multiply functions. Complete failure means that some function of the system is completely lacking, as in the example of the pocket calculator. Total lack of all function, perhaps caused by a fault in a system's power supply, would also be complete failure, but a complete failure need not be total. A fault in a digital voltmeter could prevent measurement of voltage on one range but not on others. This would constitute a partial failure of the voltage measurement function. A more subtle failure is also possible in this case: suppose the voltmeter is accurate to ±1% when working correctly and a gradual fault occurs such that the voltmeter reading is in error by 5%. The voltmeter has failed even though it is still able to measure voltage. This type of failure is known as a drift failure and can occur in measuring systems and instruments wherever some performance parameter can drift over time. If the drift is consistently in one direction it is said to be monotonic, and once the performance drifts outside an acceptable range the drift failure is permanent. If the drift is nonmonotonic, it is possible for performance to drift back into the acceptable range after a drift failure so that the failure is nonpermanent.

Failure mechanisms

Failures in electronic systems are usually attributable to failure of a component or electrical joint within the system. Component failure can be due to damage or stress during system assembly. A component might be overheated during soldering, for example, or damaged by electrostatic discharge during handling. This type of failure is known as infant mortality, and it occurs soon after system assembly. Component failure can also be due to an inherent weakness or flaw caused by defective materials used in manufacture of the component or by abnormalities in the manufacturing process. This type of failure is known as a freak failure, and it can occur after thousands of hours of use. A third possible cause of component failure is misuse: a component may be subjected to stress beyond its normal ratings. This could be the consequence of a design error where the designer failed to consider behaviour of the system under all relevant conditions, or could be a secondary failure resulting from failure of another component in the system. A short-circuit fault in a power-supply reservoir capacitor, for example, could cause secondary damage to the bridge rectifier before the fault current was interrupted by a fuse.

A final failure mechanism is known as wear-out, and is due to component ageing or wear. Wear-out failures become more likely as a system becomes older. Ageing is due to physical and chemical processes that cause deterioration of a component over time, whereas wear occurs mainly in components with moving parts such as switches and potentiometers where mechanical abrasion or sparking gradually removes material from the contacting surfaces.

The "bathtub" curve

Consider a population of electronic products in mass production. If the number of failures per unit time, or failure rate, is plotted against time from manufacture, a curve resembling that shown in Figure 9.1 is often found, and is known from its shape as the bathtub curve. There are three regions shown in the figure. In the early failure region there is initially a high failure rate due to infant mortality failures. As the number of infant mortalities falls, typically after 10 hours of operation, the failure rate rises again and then falls, due to freak failures that continue to occur at a declining rate up to 100 or 1000 hours of operation. The constant-failure-rate region normally lasts for five to ten years, during which a low incidence of freak failures occurs. This region is often referred to as the useful life or service life of the product. The total number of failures in these first two regions is normally a small fraction of the total population. Failures in these regions are normally repaired (assuming the product is accessible and designed for repair) and then continue to function throughout the constant-failure-rate region. The final part of the bathtub curve is known as the wear-out region. During this period, failure rate increases as components begin to fail through ageing and

The classical bathtub curve illustrated here is given in many texts but is not typical of electronic systems because it does not distinguish between infant mortality and freak failures.

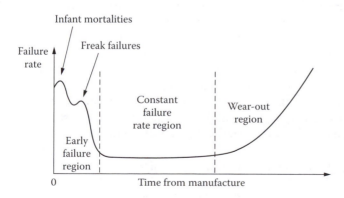

Figure 9.1   The "bathtub" curve.

155

wear. Repairs may still be effected to keep a unit serviceable, but eventually the cost of continued repair becomes uneconomic and the unit has to be withdrawn from service. All engineering products suffer this fate eventually. In the case of modern consumer products, repair may be uneconomic (it is almost as cheap, or cheaper, to buy a new unit), but in many commercial, industrial, and military applications, repair is essential to keep a larger system operational.

The infant mortality failure rate can be reduced by quality control in manufacturing and assembly and by inspection and testing of components before assembly. Poor-quality soldered joints, for example, might be avoided by regular inspection of the soldering machine or by testing sample boards run through the machine at intervals during a production run. The added cost of quality control must be balanced against the savings due to reduced incidence of failures and improved production yield.

If the products are shipped out of the factory directly after testing, the user or customer may experience early failures due to infant mortality. The added cost to the manufacturer of repairing or replacing the failed units can be avoided by burn-in testing: all units are tested under operating conditions in the factory for long enough to catch most of the early failures before they reach the customer. Mild stress such as heat, vibration, and so on can increase the proportion of weak units detected in a given time, although there may be some reduction in the service life of the whole population as a result of the stresses. Detection of weak units in this way is known as stress screening.

Stress screening is an expensive process and is normally applied only to military and aerospace electronic systems. This subject is discussed further in Chapter 10.

The failure rate during the useful life of a product where the failure rate is fairly constant can be reduced by several means. One is to use high-reliability components that have been burned in by the component manufacturer so that early freak failures have been eliminated; another is to derate components, operating them well within their normal ratings. Resistors, for example, might be operated at no more than 50% of their normal power rating and at least 40°C below their normal maximum operating temperature. Diodes and transistors might be derated to 50% of their normal anode and collector currents respectively, and all semiconductor devices might be limited to junction temperatures below 90°C (silicon). Further reductions in system failure rate can be achieved by redundancy and by minimizing the number of components for reasons, which will be seen later in this chapter.

Wear-out failures can be postponed, but not prevented, by replacing worn or aged components before they fail. This is known as preventive maintenance and extends the useful life of a product. Some components may be replaced only if worn or shown to be aged by testing, but, in many cases, components may be replaced after a fixed time or amount of use. On a moderate-sized electronic system that is not operated continuously, an elapsed time indicator or hour meter can be a useful and inexpensive way of recording actual operating time so that preventive maintenance can be carried out at the correct time.

Measures of reliability and maintainability

Reliability is a quantitative subject, and there are several quantitative measures of reliability.

Maintainability is not easily quantified, but the mean time taken to repair a faulty system can sometimes be a useful measure of maintainability.

Mean time between failures

During the useful life of an electronic product, the failure rate, usually denoted by λ, is approximately constant and is a useful measure of a system's reliability. The reciprocal of the failure rate is known as the mean time between failures (MTBF):

$$MTBF = \frac{1}{\lambda} \tag{9.1}$$

which is usually stated in hours. As an example of the level of reliability achievable with modern components, linear power supplies of up to 200 W rating can have MTBFs of more than 300,000 hours or 34 years. The corresponding failure rate is 3.3×10^{-6} hour^{-1}.

For nonrepairable systems, the mean time to failure, or MTTF, is quoted. In the case just given where the MTBF is 34 years, it is unlikely that the system will operate for another 34 years on average after a repair because of component ageing, and the figure given should properly be quoted as an MTTF.

MTTF and MTBF can also be defined in terms of the mean lifetime of a population of units and the mean interval between faults occurring in a population of repairable systems. If a sample of *n* units is tested or operated until all have failed, and the lifetime or time to failure, τ, is recorded for each unit, the MTTF is the mean lifetime.

$$MTTF = \frac{1}{n}\sum_{i=1}^{n} \tau_i \tag{9.2}$$

Equation 9.1 is valid only if the failure rate, λ, is constant.

The value given by Equation 9.2 is the observed MTTF for the population of *n* units. It can only be an estimate of the MTTF of the unit design, since the number tested can only ever be a sample of the total population.

Exercise 9.1

A sample of 10 filament lamps is tested to failure with the following lifetimes in hours: 204, 1473, 650, 697, 1737, 558, 723, 215, 526, and 1850. What is the observed MTTF?

(*Answer*: 863 hours.)

Mean time to repair

If a system is repairable, its maintainability can be quantified as the mean time to repair (MTTR). Maintainability can be influenced by design, particularly in ease of access to components; the method of repair, and the ease with which a fault can be diagnosed. Apart from design, MTTR can also be influenced by other factors such as geographical location of the system: the MTTR for a mobile or cellular phone base station located on a remote hilltop will be longer than that for the same design of base station located near a major road intersection.

A system might be repaired by component replacement or by module replacement.

Availability

In some applications, high reliability is essential either because human life depends on continued operation or because repair is impossible or very expensive. In other applications, however, where repair is possible, a more useful quantity for comparing one system with another is availability, defined as the proportion of time during which a system was, or is likely to be, functional.

Worked Example 9.1

Compare the availability figures for two computer systems (for example, mainframes used for air traffic control). System 1 has a stated MTBF of 5000 hours and can be repaired by a service technician in 8 hours on average. System 2 has an MTBF of 2000 hours and can be serviced in 2 hours on average (by swapping modules).

Solution   The availability, A, can be calculated as

$$A = \frac{\text{MTBF}}{\text{MTBF} + \text{MTTR}}$$

Hence, for system 1,

$$A = \frac{5000}{5000 + 8} = 99.84\%$$

and for system 2,

$$A = \frac{2000}{2000 + 2} = 99.90\%$$

Thus, although system 2 is likely to fail more often, its availability is marginally (but not significantly) better.

Note: In a practical situation, the consequences of failure such as loss of data and costs of repair would also have to be evaluated.

Reliability

MTBF is one quantitative measure of reliability, applicable where failure rates are constant. Another quantity, of more general applicability, is denoted by the word *reliability* itself. This is the probability that a device or system will function without failure over a specified time period or amount of use under specified operating conditions. For some devices and systems reliability is expressed per unit of time, whereas for other items reliability is stated per unit of use. A filament lamp, for example, might have an expected life of 1000 hours: the failure mechanism of the filament is dependent mainly on the time that the filament has been incandescent. A switch, on the other hand, could have a life stated as a number of operations because the failure mechanism is dependent on wear of the switch contacts.

In general, reliability, in the sense just defined, varies over time and is stated as a reliability function, $R(t)$, which gives the probability that an item will function without failure over time, t. The value of $R(t)$ can be found, in principle, from life trials on a sample of items sufficiently large to give statistically valid results. The sample of items is tested to failure, and the proportion of failed items is plotted as a function of time. The result is called the lifetime distribution function, or $F(t)$. The slope of $F(t)$ is never negative since the proportion of failed items cannot decrease. $F(t)$ can also be interpreted as the proportion of items with lifetimes less than or equal to t. Then $1 - F(t)$ is the proportion of items with lifetimes exceeding t, which is also the probability that an item will function without failure over time t, or $R(t)$. Thus,

$$R(t) = 1 - F(t) \tag{9.3}$$

The slope of $R(t)$ is zero or negative for all values of t, and for sufficiently large values of t, $R(t)$ becomes zero, because no item can continue to function indefinitely. These statements follow from Equation 9.3, and they can also be justified from the definition of $R(t)$.

It is assumed that all faults are permanent. This need not be so: faults that appear and then disappear are referred to as intermittent.

In practice, life test data are often analysed by plotting $1/R(t)$ on a double logarithmic scale [ln ln $1/R(t)$] against lifetime on a logarithmic scale. This technique is known as Weibull analysis and is discussed by O'Connor (2002).

Exercise 9.2

Explain why the slope of $R(t)$ cannot be positive, using only the definition that $R(t)$ is the probability that a unit will function without failure over a time period, t.

Derivation of failure rate

The failure rate, λ, which in general varies with time, can be derived from the reliability function $R(t)$ introduced above, making no assumptions about the form of $R(t)$ and $\lambda(t)$ other than those stated above, namely, that $R(t)$ must have zero or negative slope for all values of t.

Consider N items on test and represent the number of items failed as a function of time by $n(t)$. From the definition of the lifetime distribution function as the proportion of failed items,

$$F(t) = \frac{n(t)}{N} \tag{9.4}$$

The rate of failure of each item is the rate at which items fail divided by the number of items still functioning at time t, which is $N - n(t)$.

$$\lambda(t) = \frac{1}{N - n(t)} \frac{dn(t)}{dt} \tag{9.5}$$

From Equations 9.3 and 9.4,

$$n(t) = N(1 - R(t)) \tag{9.6}$$

and, differentiating with respect to t,

$$\frac{dn(t)}{dt} = -N \frac{dR(t)}{dt} \tag{9.7}$$

159

Substituting Equation 9.7 into Equation 9.5,

$$\lambda(t) = \frac{-N}{N - n(t)} \frac{dR(t)}{dt} \qquad (9.8)$$

but from Equation 9.6,

$$\frac{-N}{N - n(t)} = -\frac{1}{R(t)} \qquad (9.9)$$

and substituting from Equation 9.9 into Equation 9.8,

$$\lambda(t) = -\frac{1}{R(t)} \frac{dR(t)}{dt} \qquad (9.10)$$

Equation 9.10 is a general relationship between failure rate and the reliability function, $R(t)$.

Exercise 9.3

Show that an exponentially decreasing reliability function, $R(t) = \exp(-kt)$, where k is a constant, corresponds to a constant failure rate, λ, equal to k.

A useful approximate relationship between $\lambda(t)$ and $F(t)$ can be derived from Equations 9.3 and 9.10 for small values of $F(t)$. Differentiating Equation 9.3 with respect to t,

$$\frac{dR(t)}{dt} = -\frac{dF(t)}{dt} \qquad (9.11)$$

and rewriting Equation 9.10 in terms of $F(t)$,

$$\lambda(t) = \frac{1}{1 - F(t)} \frac{dF(t)}{dt} \qquad (9.12)$$

Now, if $F(t) \ll 1$ [say, $F(t) < 0.1$], then Equation 9.12 can be approximated by

$$\lambda(t) = \frac{dF(t)}{dt} \qquad (9.13)$$

and the failure rate is approximately equal to the gradient of $F(t)$.

Worked Example 9.2

A batch of 5000 road vehicle electronic ignition units was monitored during their first year in service. The number of failures per month was as follows:

Month:	1	2	3	4	5	6	7	8	9	10	11	12
Failures:	0	1	0	9	9	3	8	28	15	3	7	2

Plot the lifetime distribution function and use Equation 9.13 to produce a plot of failure rate.

The cumulative number of failures each month and the proportion of failed units, $F(t)$, are:

Month:	1	2	3	4	5	6	7	8	9	10	11	12
Cumulative failures:	0	1	1	10	19	22	30	58	73	76	83	85
$F(t)$ (%)	0	0.02	0.02	0.20	0.38	0.44	0.60	1.16	1.46	1.52	1.66	1.70

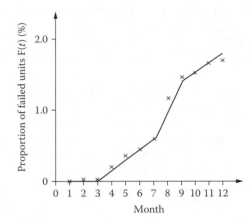

Figure 9.2   Plot of $F(t)$ for the data in Worked Example 9.2.

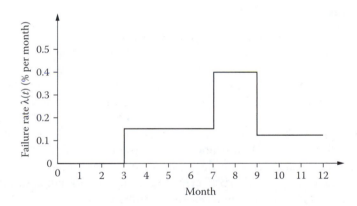

Figure 9.3   Plot of failure rate $\lambda(t)$ for the data in Worked Example 9.2.

The proportion of failed units, $F(t)$, is plotted in Figure 9.2. Approximate slopes have been drawn, and from these the failure rate, $\lambda(t)$, has been plotted in Figure 9.3 using the approximation of Equation 9.13. These are all early life (freak) failures — a domestic road vehicle is typically not used continuously, and the normal service life region of the bathtub curve is not reached until the second year of use or later.

Typically a domestic road vehicle averages about 20,000 km per year. Assuming an average speed of 60 kilometres per hour (kph), this represents about 330 hours of running time. The distinction between electronic systems that run continuously and those that are used intermittently is an important one from a reliability viewpoint.

Exercise 9.4

Plot the failure rate data for Worked Example 9.2 directly from the number of failures per month, and compare the result with Figure 9.3.

High-reliability systems

In some applications of electronic systems, high levels of reliability, or low probabilities of failure, are required. Some examples are control systems in petrochemical plants and nuclear power stations, railway signalling systems, instruments and control systems on board aircraft, and satellite communication systems. In some of these examples, high reliability is required because human life depends on correct functioning of the electronic system; in others, the consequences of failure are financial losses on a huge scale: a geostationary communications satellite cannot be salvaged, and a failure can only be remedied by launching a replacement satellite. Many applications of electronics are possible only because of the inherently low failure rate of electronic components, particularly integrated circuits (ICs), which are functionally complex but are almost as reliable as a single transistor. The design of high-reliability systems requires an understanding of the dependence of a system's reliability on the reliabilities of its subsystems and their components. This is achieved by analysis of reliability models that represent the overall reliability of a system in terms of the reliability of its components and subsystems.

Margin note: Satellites in low Earth orbit have been recovered by the NASA Space Shuttle. The cost of the failure is still enormous. An excellent example of a successful high-reliability system is the pair of Voyager space probes that took over ten years to travel to the outer planets of the Solar System. A brief article on the reliability aspects of the two Voyager probes appeared in *IEEE Spectrum* 18 (10) (October 1981): 68–70, and although the technology used is now very dated, the principles of highly reliable design remain the same.

Series systems

The simplest reliability models are of systems where every component must be functional for the system to be functional: failure of any component implies failure of the system. Many electronic systems with no special provisions for high reliability are of this type: every component has a functional purpose, and all are essential if the system is to work correctly. Figure 9.4 is a diagram of the reliability model for such a system. For reliability purposes, each component is in series with all the other components. As a simple nonelectronic example of such a system, consider a suspension bridge: the two main cables from which the bridge deck is suspended are a series reliability system since failure of either cable would cause collapse of the bridge deck.

The overall reliability of a series system is the product of the reliabilities of the subsystems. For the system shown in Figure 9.4, if the reliability of subsystem 1 is R_1, 2 is R_2, and 3 is R_3, then the reliability of the system as a whole is $R_1 \times R_2 \times R_3$. In general, when a series system has many more than three subsystems with differing individual reliabilities, the overall reliability of the system is dominated by the reliability of the least reliable

Margin note: R_1, R_2, and R_3 are probabilities that the subsystems will continue to function over a specified time period or amount of use. The overall reliability is the compound probability that all three subsystems will continue to function. Probability is discussed by Cluley (1981).

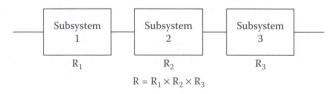

Figure 9.4   A series reliability model.

162

subsystem. The reliability of a series system is generally better if the system has as few components as are necessary for the purpose.

Exercise 9.5

Show that the reliability of a series system cannot exceed the reliability of the least reliable component.

Parallel systems

In contrast to the series reliability system just discussed, in a parallel system such as that shown in Figure 9.5, not all of the parallel units need be functional for the system as a whole to be functional. Many road vehicles are now equipped with dual-circuit braking systems and can be brought safely to a halt even if one of the brake circuits has been damaged. Commercial airliners are able to fly with only two out of four, or one out of two, engines working, and the control systems for ailerons, flaps, and rudder are at least duplicated and possibly triplicated to achieve the high level of reliability necessary for safety.

In a simple parallel system, any one of the subsystems must be functioning for the system as a whole to be functioning. This is called a 1-out-of-n system. If we use the probability of failure, F, of each subsystem, where $F = 1 - R$, then the probability of failure of the total system is the probability that all n parallel subsystems have failed. This is given by the product of the probabilities of failure of the subsystems. Qualitatively, the overall reliability of a 1-out-of-n system depends on the reliability of the most reliable subsystem. Often, of course, the parallel subsystems are identical.

More elaborate parallel systems are often used in practice, and Figure 9.6 shows a scheme known as triplicated modular redundancy (TMR). (This is a functional block diagram of the system, not a reliability model.) A system of this type might be used in a nuclear power station as a trip system for a reactor: the system input could be reactor temperature, and the system output a signal to shut down the reactor if the temperature was detected to be too high.

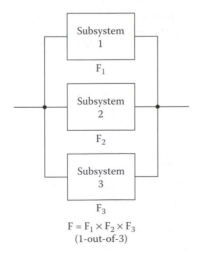

$$F = F_1 \times F_2 \times F_3$$
(1-out-of-3)

Figure 9.5   A parallel reliability model.

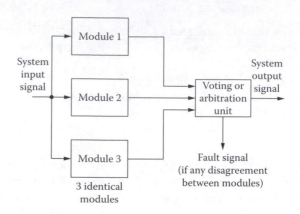

Figure 9.6   A triplicated 2-out-of-3 system.

In a triplicated modular redundancy system, three identical electronic units or modules perform some function using a system input signal. All three modules normally produce identical outputs, but the system is designed to work correctly even if one of the modules fails and produces an output differing from that of the other two modules. A TMR system is thus a 2-out-of-3 parallel system. A simple, highly reliable voting or arbitration unit produces a system output based on agreement between any two of the three modules and also generates a fault signal if the three modules do not agree. Normally the voting unit is designed to be fail-safe. In the example of the nuclear reactor trip system, the voting unit would be designed to signal a reactor shutdown if no two modules agreed or if the voting unit itself was faulty.

Worked Example 9.3

Derive the reliability, R, of the 2-out-of-3 TMR system shown in Figure 9.6.

Solution   Let the reliability of the voting unit be R_v, the reliability of the triplicated modules be R_m, and the corresponding probability of failure of a triplicated module be F_m, where $F_m = 1 - R_m$.

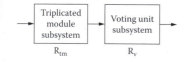

The system can be modelled as a series system of two elements, as sketched in the margin (this is a reliability model). The overall reliability of the TMR system using the rule for a series system is

$$R = R_{tm} \times R_v$$

where R_{tm} is the reliability of the triplicated module subsystem, R_{tm} is the probability that at least two of the redundant modules are functioning, and $F_{tm} = 1 - R_{tm}$ is therefore the probability that two or all three of the redundant modules fail. There are three ways in which two out of the three modules can fail and one way in which all three can fail:

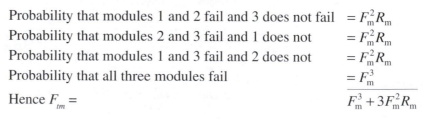

Probability that modules 1 and 2 fail and 3 does not fail $= F_m^2 R_m$

Probability that modules 2 and 3 fail and 1 does not $= F_m^2 R_m$

Probability that modules 1 and 3 fail and 2 does not $= F_m^2 R_m$

Probability that all three modules fail $= F_m^3$

Hence $F_{tm} =$ $F_m^3 + 3F_m^2 R_m$

Since $R_{tm} = 1 - F_{tm}$, the overall reliability of the TMR system is

$$\left(1 - F_m^3 + 3F_m^2 R_m\right)R_v$$

which can be shown to be

$$\left(3R_m^3 - 2R_m^2\right)R_v$$

Exercise 9.6

Show that $1 - F_m^3 - 3F_m^2 R_m = 3R_m^2 - 2R_m^3$.

Worked Example 9.4

Compare the reliability of a TMR measurement system with a nonredundant system using the same measuring module design if the module reliability is 0.999 and the voting unit is 10 times more reliable than the measuring module.

Solution The voting unit has a reliability, R_v, of 0.9999 because its probability of failure, F_v, must be 10 times less than the module probability of failure, F_m, which is 0.001. Therefore the reliability of the TMR system from Worked Example 9.3 is

$$(3(0.999)^2 - 2(0.999)^3)0.9999 = 0.99990$$

and the probability of failure is 10^{-4}. This is a factor of 10 better than the reliability of a single module.

Diversity

A possible weakness in parallel, redundant systems is the so-called common-mode failure, where all redundant units fail simultaneously from the same cause. As an example, if all the engines on a multiengined airliner were supplied with fuel from the same fuel pump, a failure of the fuel pump would stop all the engines. However, even if separate fuel pumps are provided, there is still the possibility of a common-mode failure under some unlikely operating conditions because all the pumps are of the same design and all could fail at the same time if the unlikely condition occurs. This problem is especially severe with engineering software since identical copies of a program will always give identical results no matter what the input data. In highly critical systems, therefore, there may be a need to employ redundant parallel systems of different design to each other. Such an approach can be very expensive, especially in software-based systems, where the cost of developing one system can be high enough: a diverse redundant system requires two design teams developing software for two different computers and maybe even programming in two different languages.

An alternative for small software systems is to prove the software to be correct mathematically. Much research effort has been expended on the problem of proving the correctness of both programs and complex logic systems such as microprocessors. These techniques have so far been applied only in limited but important cases, such as aircraft and railway signalling (e.g., the French Transmission Voie-Machine (TVM) signalling system used on high-speed lines at up to 350 km per hour[1]).

Maintenance

Maintenance is defined in British Standard 4778 as

The combinations of all technical and administrative actions, including supervision actions, intended to retain an item in, or restore it to, a state in which it can perform a required function.

Sometimes the word *servicing* is used to refer to these activities, and in many organizations the department responsible for carrying out maintenance is called the servicing department. Maintenance activities include calibration (adjusting a system so that its characteristics are within a specified range), fault diagnosis (establishing the cause of a failure), repair (removing and replacing faulty, damaged, or worn parts and components), and testing (checking that a system is functioning correctly, especially after repair). The design of an electronic system can have a significant effect on the cost of performing these activities, and often quite simple changes at the design stage can greatly reduce the cost of maintenance.

Maintenance strategies

Some electronic products are designed for nil maintenance because of inaccessibility once in service or because it is cheaper to scrap a failed unit than to repair it.

There are many different approaches to the maintenance of an electronic product dependent on the nature and value of the product, the type of manufacturer, the type of user, and the desired level of reliability. The first decision to be made about a product is whether it is to be maintained at all. Some products are inherently nonmaintainable; others are not economically maintainable. As examples, an IC cannot be repaired and it is unlikely that a cheap digital wristwatch can be repaired for less than the original cost. Secondly, if the product is to be maintained, there is a choice between maintenance in the field (at the user's premises) or by returning the product to a service centre. In some cases, the choice will be determined by the nature of the product and the type of maintenance needed. Large composite systems may be maintained by a combination of the two approaches, with major subassemblies maintained in the field and smaller units returned to a service centre for maintenance. Finally, there is a choice between preventive maintenance or the replacement of items before they wear out and corrective maintenance as needed to repair faults. Preventive maintenance can be applied only where components are liable to fail through wear-out. The majority of electronic components, especially semiconductors, have a useful life far in excess of the likely life of a system, so that failures due to these components are rare, per component, and are randomly distributed in time. Preventive maintenance cannot therefore reduce the failure rate attributable to these components.

Fault diagnosis

An important and often difficult part of the process of repairing an electronic system after a failure is that of diagnosing the fault, or establishing what repair action is necessary to restore the system to its normal functioning state. Some faults, by their nature, are obvious and require little diagnosis. A failed indicator lamp, for example, can be seen to have failed and can be replaced. Most faults in electronic equipment, however, are much more subtle, often manifesting themselves by obscure symptoms. These types of fault often require considerable skill to diagnose and may require the use of expensive and sophisticated test equipment, such as logic analysis. Computers, being among the most complex of electronic systems, often present difficult problems in fault diagnosis.

If an electronic system is sufficiently complex to have been designed in modular form, the first step in diagnosing a fault is to establish which module is faulty. This can be done by observing the symptoms of the fault, or by deduction from the observed behaviour. The suspected faulty module can then be replaced with an identical spare, and the system tested to confirm whether the fault has been eliminated. If not, the original module can be replaced and a different possibility investigated. This stage of fault diagnosis by modular replacement can be carried out by someone with little knowledge of the detailed workings of each module. One problem, however, is that spare modules are needed for substitution, unless the system contains more than one of each module type.

Diagnosis of faults to electronic component level is more difficult than isolating a fault to a single module, since a detailed understanding of the circuit is necessary, together with skill and experience in fault-finding. The designer of a board can make fault-finding easier by adding test points or special terminals to allow easy connection of test equipment. Also, if components are mounted in sockets, faults can be diagnosed by swapping suspect components. Many other possibilities exist for aiding testability and fault diagnosis by design.

See Wilkins (1990) for examples in the field of logic design.

Design for maintainability

Attention must be given to the maintainability of a product at the design stage if maintenance costs are to be kept low. Clearly, if a product is to be designed for nil maintenance, construction techniques can be used that would not be acceptable on a maintained product. IC sockets, for example, are not worth using if the PCB on which they are mounted is to be a throw-away, nonrepaired item. Assuming that we are considering only maintainable products here, what steps can be taken in design to make maintenance easier?

Perhaps the most important aspect of product design for maintainability is access: how easy is it to get at those parts of the product that require replacement, adjustment, or inspection? Covers and panels should be easily removable and preferably free of attached wiring. Printed circuit boards should have connectors rather than permanently soldered wiring to simplify removal and replacement. Wiring to switches and so on can be fabricated with crimped push-on connectors, both to simplify manufacture and to make switch replacement easier.

Adjustable components such as preset potentiometers and trimmer capacitors should be used as little as possible, because of the time taken to set them during manufacture and because of the need to adjust them during maintenance. Where they cannot be avoided because the required circuit properties cannot be achieved with fixed components, they should be readily accessible and capable of adjustment while the equipment is operating. A test point or output should be provided so that the effect of the adjustment can be monitored.

Summary

Electronic systems have a limited life due to deterioration of components and materials with time. Components can fail due to inherent weakness, misuse, and wear-out, and can cause partial or complete failure of the system of which they are a part. The bathtub curve represents a typical pattern of failure rates for electronic systems characterized by a relatively high early failure rate as weak components fail (infant mortality and freak failures), a low fairly constant failure rate thereafter, and eventually an increasing failure rate due to component wear-out. The bathtub curve explains why electronic products are often tested for a period after manufacture (burn-in) and how the operating life of a system can be extended by replacing components before they fail due to wear-out.

Mean time between failures is a common measure of reliability and is the reciprocal of failure rate, λ, provided λ is constant. A general reliability function, $R(t)$, can be defined even if failure rate varies with time. $R(t)$ is the probability that an item will be functioning at time, t. $R(t)$ can be measured by determining the lifetime distribution function, $F(t)$, from a life test on a sample of items. $R(t)$ is $1 - F(t)$. The failure rate, $\lambda(t)$, can be expressed in terms of $R(t)$.

Where high reliability is needed, parallel redundancy can be applied to overcome the limited reliability of systems where every component must function for the system as a whole to function. Reliability models can be used to predict the reliability of redundant and nonredundant systems using simple probability methods. To overcome the problem of common-mode failure, diverse redundant systems of different design can be used.

Maintenance activities include calibration, fault diagnosis, repair, and testing. Attention to maintainability at the design stage is essential if unnecessary maintenance costs are to be avoided.

Problems

9.1 Over a 6-month period, a computer installation suffered four failures, taking respectively 1, 2, 1.5, and 2 hours to repair. Also the system was shut down on two occasions for routine preventive maintenance for 2 hours each time. What are (a) the availability and (b) the MTBF over this period?

See Exercise 9.3.

9.2 A designer requires a power supply with a reliability of 0.8 over 2 years. (a) Assuming constant failure rate, what MTBF is required? (b) The best suitable power supply available has an MTBF of 50,000 hours; if two of these are used in parallel with a number of diodes to isolate a failed unit, what reliability must the diodes have?

9.3 An instrument system for a chemical plant has two redundant measuring channels with an MTBF of 100,000 hours sharing a power supply with an MTBF of 150,000 hours. What is the probability of a system failure causing loss of the measurement function within 5 years?

9.4 An electronic unit has an MTBF of 100,000 hours and operates continuously. Assuming a constant failure rate

$$[R(t) = \exp(-\lambda t)]$$

calculate the age at which this unit should be replaced if its reliability is not to fall below 75%.

Environmental factors and testing   10

Objectives

□ To discuss a selection of common environmental factors that influence the performance and reliability of electronic systems.
□ To discuss type testing and electronic production testing.
□ To introduce design for testability.

Environmental factors

This book opened with the statement that modern electronic products are to be found in a wide range of applications environments. These products must not only be fit for their intended purpose but also be able to survive the stresses imposed by their operating environments during their design lives. Some environments are relatively benign and impose little stress on an electronic system. A domestic living room, for example, provides a controlled-temperature, low-humidity environment reasonably free from mechanical stress and electromagnetic fields. Other environments, such as a road vehicle engine compartment, are much more hostile and subject to an electronic system to wide variations in temperature, high humidity, mechanical shock, vibration, and grime. The effects of many environmental factors are not easy to predict, and indeed the presence of the factors themselves may not be foreseen at the design stage. There is thus a need for testing in simulated environmental conditions and, in some cases, full field trials in the actual operating environment to establish that the product will work as intended.

We can now discuss a selection of common environmental factors.

Temperature

The temperature of the operating environment is an obvious factor to be considered during design. There are three likely temperature ranges over which an electronic product might be required to operate, as listed in Table 10.1. These are broad generalizations, and in most applications some variation of these ranges might be encountered.

Circuit design for a wide operating temperature range is made difficult by the temperature dependence of many component parameters.

Table 10.1   Operating temperature ranges for electronic equipment

Designation	Range (°C)	Environments
Commercial	0 to +70	Homes, offices, and laboratories
Industrial	−40 to +85	Process plants, factory floors
Military	−55 to +125	Weapons systems, fighting vehicles

Examples include the common emitter current gain, β or h_{fe}, of bipolar transistors; the input offset voltages of operational amplifiers; the breakdown voltages of Zener diodes; and, of course, the values of resistors and capacitors, which were discussed in Chapter 5.

Worked Example 10.1

A weighing machine is required to weigh 0 to 25 kg in 10 g steps and to give a numerical readout. What is the precision needed in the machine's analogue-to-digital converter (ADC)? The designer decides to use a 5.1 V BZX79 Zener diode to provide the voltage reference for the analogue-to-digital converter. The diode has a temperature coefficient of breakdown voltage between −2.7 and +1.2 mV°C^{-1}. If the operating temperature range is +5°C to +40°C, what would be the contribution from the Zener voltage reference to the worst-case error in the machine's reading if it was calibrated at 20°C?

Solution   10 g in 25 kg is 1 part in 2500, or 0.04%. (An ADC of 12-bit precision would be needed.) The worst-case temperature is 40°C, or a change of 20°C from the calibration temperature. Assuming the worst-case temperature coefficient of −2.7 mV°C^{-1}, the Zener diode voltage will change by −2.7 × 20 mV, or −54 mV from its 20°C voltage or a change of 1%. The error caused by the temperature variation of the Zener diode voltage is thus over 20 times greater than the precision required (and all other factors contributing to the error have been ignored). Hence either the specification has to be relaxed or an improved voltage reference used.

Temperature variation of component parameters can be a significant problem even for systems of modest performance operating over the commercial temperature range. Many circuit techniques have been devised to reduce the effect of temperature by adding compensating components. In the case of crystal oscillators, which are often used as precision frequency references, the resonant frequency of the crystal varies with temperature. To overcome this problem, the oscillator may be housed in a small temperature-controlled enclosure or "oven" in order to reduce the variation of output frequency with ambient temperature.

Clearly, all components within an electronic system must have an operating temperature range covering at least the operating temperature range of the product itself, unless some controlled-temperature environment is provided within the system.

Particular problems exist with the selection of components for operation at subzero temperatures. At these temperatures, common difficulties include some battery types, for example, lead acid, which do not work well below 0°C, and transistor–transistor logic (TTL) integrated circuits (ICs), which are rated down to only 0°C.

Battery behaviour at low temperatures was discussed in Chapter 4. Early liquid crystal displays (LCDs) had a slow response and exhibited poor contrast below 5°C. Improved materials later reduced the minimum operating temperature for most LCDs to around −30°C.

Exercise 10.1

TTL 74 series logic has an operating temperature range of 0 to 70°C, yet complementary metal-oxide semiconductor (CMOS) logic can operate at temperatures as low as −40°C. Why?

A less obvious, but very important, effect of temperature is a reduction in the reliability of a product when operating at an elevated temperature or when subjected to repeated variation in temperature (thermal cycling). One way in which temperature and temperature cycling reduce reliability is by the creation of mechanical stresses within components and joints by differential thermal expansion, which may lead to fracture. This problem occurs with ceramic chip carriers soldered to epoxy–fibreglass printed circuit boards (PCBs), for example, where the temperature coefficient of expansion (TCE) of the ceramic (about 6 parts per million [p.p.m.] °C^{-1}) and the epoxy–fibreglass (about 12 to 16 p.p.m. °C^{-1}) are sufficiently different to make this method of mounting unreliable because of solder joint fracture after temperature cycling.

Possible solutions to this problem are: mounting the ceramic chip carriers in sockets, using a special-purpose PCB laminate with a TCE matched to that of the ceramic chip carrier, and mounting the chip carriers on a ceramic substrate.

Worked Example 10.2

Some of the largest chip carriers have 124 pads, 31 per side, spaced at 1.27 mm centres. The distance between pad centres at the ends of a side is 30×1.27, or about 38 mm. Over a 50°C temperature range, the relative change in end-pad spacing between the chip carrier and the board is 38 mm $\times$ 50°C $\times$ 6 to 10 p.p.m. °C^{-1}, or 11 to 19 μm. This is about 1 to 1.5% of the pad spacing of 1.27 mm.

Similar problems with differential thermal expansion occur in plastic IC packaging, both between the silicon chip and the metal leadframe to which the chip is bonded and between the metal leadframe and the moulded plastic package. The TCEs of the leadframe and the plastic have to be matched to the TCE of silicon, which is about 2.6 p.p.m. °C^{-1}, to prevent fracture of the bond between the chip and the leadframe and to reduce moisture penetration along the leads.

A second, very significant effect of temperature on reliability is an increase in the rate of component ageing or deterioration with increased temperature. Many component wear-out failures are caused by chemical reactions occurring inside the component, and because the rate of reaction increases with temperature, the life of a component is reduced by operating at higher temperatures. Many chemical reactions obey the Arrhenius equation

The Arrhenius equation is discussed by O'Connor (2002), and in its chemical context in any "A" level or high school physical chemistry text. O'Connor expressed some doubt as to whether the equation is valid for many electronic components.

$$\lambda = K \exp(-E/kt) \tag{10.1}$$

where λ is a reaction rate or failure rate, K is a constant for any particular reaction or component type, E is an activation energy for the reaction, k is Boltzmann's constant (1.38×10^{-23} J K^{-1}), and T is absolute temperature. The activation energy for a chemical reaction is the minimum energy required by the reactant atoms or molecules for the reaction to occur. Many reactions have activation energies between 0.5 and 1.5 eV, or 0.8 and 2.4×10^{-19} J.

An electron volt (eV) is the energy imparted to a unit charge (the charge on an electron) when transferred through a potential difference of 1 V. Since the electron charge is about 1.6×10^{-19}C, 1 eV is equivalent to 1.6×10^{-19} J.

Worked Example 10.3

Calculate the increase in failure rate at a temperature of 60°C (333 K) over the failure rate at 20°C (293 K), assuming the Arrhenius equation

(Equation 10.1) applies and the activation energy for the underlying chemical process is 0.8 eV, or about 1.3×10^{-19} J.

Solution   Given two temperatures T_1 and T_2 ($T_2 > T_1$), Equation 10.1 can be applied at both temperatures:

$$\lambda_2 = K \exp(-E/k\,T_2), \qquad \lambda_1 = K \exp(-E/kT_1)$$

Dividing one equation by the other eliminates K:

$$\frac{\lambda_2}{\lambda_1} = \frac{\exp(-E/kT_2)}{\exp(-E/kT_1)} = \exp\left(\frac{E}{k}\left(\frac{1}{T_1} - \frac{1}{T_2} \right) \right)$$

(10.2)

and substituting the values of E, k, T_1, and T_2 gives

$$\frac{\lambda_2}{\lambda_1} = \exp\left(\frac{1.3 \times 10^{-19}}{1.38 \times 10^{-23}} \left(\frac{1}{293} - \frac{1}{333} \right) \right) = 48$$

The failure rate therefore increases by a factor of about 50 for a temperature rise of 40°C at this activation energy, if the Arrhenius equation holds.

The increase in failure rate at higher temperatures has two important consequences. First, long-term reliability tests can be speeded up by operating the devices under test at elevated temperature, yielding results in a shorter time of perhaps weeks or months rather than years, and therefore reducing the costs of testing considerably. This is called accelerated life testing. Second, product reliability is improved if components are kept cool. Power-dissipating components such as resistors and semiconductors can be derated by using a component with a higher power rating than is strictly necessary for the purpose. This derating will result in the operating temperature of the component being reduced with a consequent increase in reliability. Power semiconductors can also be operated at lower temperatures by attaching them to more effective heatsinks, or by providing forced cooling. In applications where electronic subsystems form part of a larger system, operating temperatures may be reduced by careful siting of the electronics away from sources of heat.

Exercise 10.2

A capacitor manufacturer quotes lifetimes for a range of capacitors as 120,000 hours at 40°C and 6000 hours at 85°C. If the failure rate is assumed constant and therefore proportional to the reciprocal of the component lifetime, what is the activation energy, E, and what would the lifetime be at 100°C by extrapolation?

(*Answer*: 0.64 eV, 2600 hours.)

Water and humidity

Water is an obvious environmental hazard to any electronic equipment intended to operate out of doors or in a wet industrial environment. As all water except the very purest is conductive, short circuits, leakage

O'Connor (2002) suggested that if the Arrhenius equation is in fact pessimistic, many electronic components are overcooled or derated more than necessary, and could be operated reliably at higher temperatures or power dissipations.

A single activation energy is assumed here for the underlying chemical process.

Table 10.2 Degree of protection against water indicated by the second characteristic numeral (of the IP classification)

Second characteristic numeral	Degree of protection	
	Brief description	Definition
0	Nonprotected	
1	Protected against vertically falling water drops.	Vertically falling drops shall have no harmful effects.
2	Protected against vertically falling water drops when enclosure tilted up to 15°.	Vertically falling drops shall have no harmful effects when the enclosure is tilted at any angle up to 15° on either side of the vertical.
3	Protected against spraying water.	Water sprayed at an angle up to 60° either side of the vertical shall have no harmful effects.
4	Protected against splashing water.	Water splashed against the enclosure from any direction shall have no harmful effects.
5	Protected against water jets.	Water projected in jets against the enclosure from any direction shall have no harmful effects.
6	Protected against powerful water jets.	Water projected in powerful jets against the enclosure from any direction shall have no harmful effects.
7	Protected against the effects of temporary immersion in water.	Ingress of water in quantities causing harmful effects shall not be possible when the enclosure is temporarily immersed in water under standardized conditions of pressure and time.
8	Protected against the effects of continuous immersion in water.	Ingress of water in quantities causing harmful effects shall not be possible when the enclosure is continuously immersed in water under conditions that shall be agreed on between manufacturer and user but that are more severe than for numeral 7.

Adapted from BS EN 60529 (1992) with the permission of the British Standards Institution.

currents, and high-voltage flashover will occur if water comes into contact with electrical or electronic circuits. There is thus a need to design equipment enclosures to prevent the ingress of water. British Standards/ European Norm (BS EN) 60529 (IEC 60529) defines nine categories of protection against water ranging from 0 (no protection) to 8 (protected against submersion), and including varying levels of dripping and spraying water in between (Table 10.2). There are standardized empirical tests defined for each category. Ideally, of course, every equipment enclosure for use in anything except a dry environment would be watertight, but totally sealed enclosures are expensive and may create problems with heat management and accessibility for maintenance. In practice, therefore, an equipment enclosure will be designed with the minimum degree of protection against water necessary for the application environment.

Humidity, or water vapour, is a less well-defined problem for the equipment designer. Humidity can be absorbed by some electronic engineering materials such as PCB laminates, causing reduced insulation resistance and increased leakage currents, which can be a problem in high-impedance circuits such as charge-sensitive amplifiers and sample-and-hold circuits.

The quantity of water vapour present in the air is given by the relative humidity (RH), which is the ratio of the density of water vapour in the air to the density of saturated water vapour in the air at the same temperature. RH can be up to 100%, at which point water condenses out as droplets forming fog or dew. Protection against water vapour under these conditions must consist of either a heated or hermetically sealed (airtight) enclosure.

On a longer time scale, water can cause corrosion, which can initiate early failure of an electronic system. Corroded connectors, for example, may not provide a sufficiently low-impedance ground or shielding connection to be effective against electromagnetic interference. Materials must therefore be selected carefully to suit the application environment. Anodized aluminium, for example, which is a cheap, attractive, and popular material for benign indoor environments, soon oxidizes and loses its surface conductivity so that grounding connections become high-impedance joints, and are therefore ineffective.

Foreign bodies

BS EN 60529, mentioned in the previous section, also defines seven categories of protection against foreign bodies ranging in size from greater than 50 mm in diameter to dust. Each category in Table 10.3 is indicated by a numeral and can be combined with a numeral from Table 10.2 to form a two-digit code, prefixed with the letters IP. Thus, "IP45" means that an equipment enclosure so designated is protected against foreign bodies to category 4 (objects of 1 mm diameter or greater) and water to category 5 (water jets). Large foreign bodies are an obvious mechanical hazard inside an equipment enclosure, while smaller objects may cause damage if they enter switches or connectors, or if they are conductive. Similarly, dust can cause damage to exposed switch contacts and may also initiate current leakage or flashover. Measures to deal with these problems can only be empirical.

Mechanical stress

Many electronic systems have to withstand mechanical stress in the form of impacts and shocks, vibration, and acceleration. Some of these are amenable to analysis, while others can be dealt with semiempirically.

Table 10.3   Degrees of protection against solid foreign objects indicated by the first characteristic numeral (of the IP classification)

First characteristic numeral	Degree of protection	
	Brief description	Definition
0	Nonprotected.	
1	Protected against solid foreign objects of diameter 50 mm and greater.	
2	Protected against solid foreign objects of diameter 12.5 mm and greater.	Defined test probes with the requisite diameter must not penetrate the enclosure more than a specified extent.
3	Protected against solid foreign objects of diameter 2.5 mm and greater.	
4	Protected against solid foreign objects of diameter 1 mm and greater.	
5	Dust protected.	Ingress of dust is not totally prevented, but dust does not enter in sufficient quantity to interfere with satisfactory operation of the equipment.
6	Dust-tight.	

Adapted from BS EN 60529: 1992 with the permission of the British Standards Institution.

Acceleration, for example, in an aircraft or missile will be known from the performance parameters of the aircraft or missile, and the forces set up in a system by acceleration can be calculated from elementary mechanics. The effect on a component or system can then be calculated, or tested in a centrifuge.

Vibration, or periodic acceleration, however, is a much more difficult problem because vibrational energy may be coupled efficiently into components at certain frequencies by resonance, even to the extent of causing breakage by the application of excessive force to joints and fastenings. Some analysis may be possible, but testing is usually necessary to establish where breakages are likely to occur. This is likely to be a task for a vibration specialist with a background in mechanical engineering, and shows that the design and development of high-performance electronic products or systems is a multidisciplinary activity.

Quite often, of course, electronic design is not the major part of a project, and electronics engineers are themselves working as specialists within a team.

Humans

Before leaving the subject of environmental stress, it is worth pointing out that people are present in many application environments and that they can be as much of a hazard to a system (and themselves) as any of the factors so far discussed. The design of enclosures, controls, connectors, and handles should take into account the likely levels of skill and sympathy of the user, and they should be sufficiently robust to withstand the use and abuse that they are likely to receive. If an enclosure is big enough to sit or stand on, for example, then perhaps it should be designed to withstand a weight of at least 80 kg. A handheld product such as a portable phone is highly likely to be dropped from a height of about 1.5 m and ought to be designed to withstand such a drop without major damage.

Drop tests are widely used for equipment likely to be dropped. If a hard unyielding surface such as steel plate or concrete is used, a drop from even 1 m can be a severe test.

Type testing

This book has attempted to show that the design of successful electronic products and systems, while based on a large body of underlying theory, is still heavily dependent on the skill, intuition, and foresight of the design team. Many factors are not amenable to precise analysis, and some are not amenable to analysis at all. There is thus a need for some form of testing of a design before production. Type testing is carried out on a prototype or several prototypes with the aim of establishing whether the design meets its performance specification and whether it complies with relevant standards or legislation, and to discover any weaknesses or design errors so that corrective design, or development, can be undertaken before the system goes into production. On a complex design, type testing can be a very expensive and time-consuming job, especially when design changes are made as a result of testing and some tests have to be repeated to verify the behaviour of the modified design.

Type testing may be destructive.

Environmental testing

To test a system over its full range of operating temperatures, atmospheric pressures, and humidities, environmental test chambers of varying sizes and degrees of sophistication are used. The smallest may be no more than a fan-assisted oven/refrigerator, while the largest may be big enough to contain a vehicle. The number of combinations of environmental parameters that can be applied to a system under test is likely to be limited by the expense and time involved so that some estimate must be made of the worst-case combinations of static conditions. Once one starts to consider dynamic conditions, the number of possible tests becomes even greater, and some pragmatic judgement has to be exercised to determine what the product is likely to encounter in its operating environment. A few hours at 18°C and high relative humidity, followed by a fall in temperature to below the dew point, for example, would simulate conditions out of doors in a temperate zone at nightfall, and would test the product's ability to withstand condensation.

Consider, for example, simulating the operating environment of a geostationary communications satellite. In this case, the real operating environment cannot be used for testing and the expense is unavoidable.

If the operating environment is accessible, field trials can be used to test a product using the real operating environment rather than a laboratory-simulated environment. A simulated environment provides more control over the testing and gives clearer results, but can also be very expensive.

Electromagnetic testing

Electromagnetic susceptibility and emissions tests are now a requirement for all electrical and electronic products sold or manufactured in the European Union (EU), even those intended for domestic and office applications. The severity of the tests applied can vary considerably, however. A missile system to be installed on a warship, for example, would be tested rigorously over a wide range of frequencies up to and including radar frequencies and at considerable field strengths, because of the proximity of the system to powerful antennae on board the ship. A radio-telephone for use in an emergency services vehicle, on the other hand, would be given a much less rigorous test, concentrating on the frequency range likely to be encountered on the road and in towns. Powerful radio sources in the very high frequency (VHF) band, for example, such as other emergency vehicles, could be encountered, so that susceptibility to frequencies of around 100 MHz would certainly be tested.

Electromagnetic testing is carried out in electromagnetic (EM) test chambers, fully or partially lined with cones of radio-absorbing material (RAM) to absorb EM waves incident on the walls and to prevent the creation of standing waves that would distort the intended field distribution in the chamber.

The EU Electromagnetic Compatibility (EMC) Directive, referred to in Chapter 8, has made electromagnetic type testing an essential part of the design and development programme for any electrical or electronic product, and not only within the EU, since many designs are sold worldwide.

Radio-absorbing material (RAM) consists of foamed rubber or plastic loaded with carbon and ferrite powder. The acronym RAM is also used for random access memory in digital electronics and computers.

Exercise 10.3

Explain why the radio-absorbing material lining of an EM test chamber is shaped into cones.

Figure 10.1 shows a microwave oven inside an EM test chamber. The test chamber provides a controlled EM environment, excluding ambient EM fields and waves so that emissions from the system under test can be identified unambiguously, and containing powerful EM fields generated within the chamber for susceptibility testing. A considerable quantity of

Figure 10.1 A microwave oven under test for electromagnetic emissions in a test chamber lined with cones of radio-absorbing material. (Courtesy of TÜV Product Service Ltd.)

expensive test equipment is needed to detect, measure, and generate EM fields, including antennae, spectrum analysers, swept frequency generators, radio frequency (r.f.) power amplifiers, and broadband receivers. Power and signals are fed in and out of a chamber through waveguide ports or through deep tubular ports acting as waveguides below cut-off. Doors must be sealed against EM fields by sprung metal fingers to ensure low-impedance grounding of the door to the chamber walls.

Electronic production testing

Once a design is in production, other types of tests are required to verify that each unit manufactured works correctly and is free from manufacturing faults that could cause early failure. There are two main methods of testing electronic products for correct operation and a variety of tests that expose the products to stress with the aim of detecting weak units.

In-circuit testing

A common test technique for newly assembled PCBs aims to establish that every component on the PCB is correctly installed, of the right type or value, and functioning correctly. At first sight this might seem a superfluous test, but in a production environment it is very difficult to avoid assembling an occasional IC back-to-front, installing a polarized component the wrong way round, or even mixing up components so that resistors of the wrong value, for example, are assembled into a board.

This is true even when automatic component insertion is used, although errors are much more likely in hand-assembled boards.

An in-circuit test machine carries out an electrical test on each component to verify its behaviour, value, and orientation. The effect of neighbouring components is avoided by guarding or back-driving. Figure 10.2 shows a simple example of the technique. An in-circuit tester is connected to the circuit shown at A and B with the aim of checking the value of R_3. To do this, a voltage of, say, 5 V is applied at A with respect to B. The tester measures the current flowing, I, to obtain the value of R_3 from $5/I$. In the circuit shown, however, the transistor's base-emitter junction will be forward-biased by the voltage at A, and some current will flow through the base-emitter diode and R_4 to B. To overcome this, the base-emitter diode can be reverse-biased by applying a voltage at D of say 6 V. Finally, current can flow from A through R_1

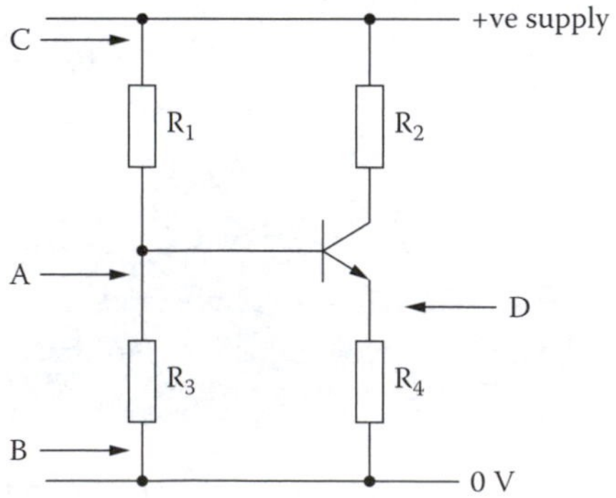

Figure 10.2 An example circuit illustrating the principle of in-circuit testing.

to the positive supply rail and through other circuits to 0 V and hence back to B. This can be prevented by applying the same voltage at C as is applied at A (from a separate source, not by connecting A and C together) so that no potential difference exists across R_1 and therefore no current flows through R_1 (or R_2 via the base-collector junction of the transistor). This is the same technique of guarding that was discussed in Chapter 8 in the context of reducing leakage currents in a sample-and-hold circuit. One problem with in-circuit testing is the possible over-stressing of neighbouring components, such as R_4 in the example above. This is normally dealt with by applying the test voltages for a short time only.

Care has to be taken as the voltage applied at D could be more than the normal voltage present when the circuit is operating, and this could overstress R_4 by dissipating a higher power than the resistor is rated to handle.

Functional testing

An in-circuit test does not check that a PCB will actually work because there are a variety of faults such as bad joints, solder bridges or splashes between tracks, and faulty plated-through holes that are not tested by an in-circuit test. A functional test checks the behaviour of the board with power applied and with test signals fed in. This type of test can be done manually using laboratory test gear such as oscilloscopes, logic analysers, and signal generators or automatically using automatic test equipment (ATE). ATE can consist of conventional laboratory test equipment fitted with interfaces such as the IEEE-488 bus and controlled by a computer (Figure 10.3a), or, for higher-volume testing, a functional tester that is a computer or microprocessor-based system containing stimulus and measurement subsystems equivalent in function to conventional test equipment (Figure 10.3b). For testing small batches of boards, the unit-under-test or UUT can be connected to the test equipment using *ad hoc* cables and connectors connected by hand. When larger quantities of boards are to be tested, a more cost-effective and faster method of connecting to the UUT is a bed-of-nails fixture, as shown in Figure 10.4. The UUT is held on the fixture by a partial vacuum, and electrical connections to the UUT are made by spring-loaded contacts or nails, touching directly onto the PCB tracks. The ends of the nails are sharp so that they cut through any surface oxide on the solder to make a low-resistance connection to the board. Electrical connections from the nails are brought out to an airtight connector and thence through a cable to the ATE system. Bed-of-nails fixtures have to be custom-made and wired for each type of UUT, but the cost can soon be recovered by the saving in time obtained by their use. Typically, a board can be tested in 1–2 minutes using functional ATE and bed-of-nails fixtures.

The IEEE-488 bus is an 8-bit parallel interface capable of working at data transfer rates of 10^6 bytes (1 byte = 8 bits) per second over distances of up to 15 m. It was developed in the 1970s but is still in use 35 years later, even though it has been replaced by USB and FireWire for office applications.

Design for testability

The use of ATE for product testing is cheaper and easier if the designer considers testability at the design stage. Electronics designers should be aware that they must consider how their design will be tested right at the start of the design process, rather than leaving test problems to be resolved when the design is almost complete.

Design for testability can consist of very simple and mundane measures such as ensuring that there are suitable places on the underside of a PCB where contact can be made by a bed-of-nails fixture, to more

Wilkins (1990) discussed design for testability extensively in the field of digital circuits.

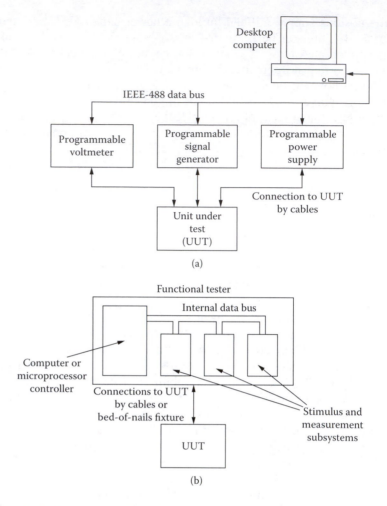

Figure 10.3   Functional ATE: (a) programmable instruments controlled by a desktop computer via an IEEE-488 interface bus and (b) a functional tester.

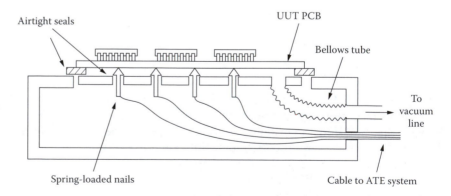

Figure 10.4   Main principles of a bed-of-nails test fixture.

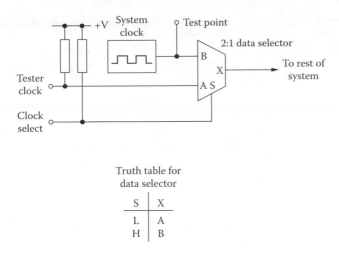

Truth table for
data selector

S	X
L	A
H	B

Figure 10.5   An example of a design for test: the clock circuit of a synchronous digital system.

elaborate circuit design techniques to make testing easier. Figure 10.5 shows an example of circuit design for testability. Synchronous digital circuits are controlled by a clock or oscillator. Testing of a synchronous circuit normally requires that the circuit be clocked by the ATE system. The figure shows how, with the addition of a 2:1 data selector and two resistors, the testability of the circuit can be improved. In normal operation the clock select input and the tester clock input are left unconnected, and the system clock input to the data selector is output to the rest of the system. When connected to the ATE system, the ATE clock is connected to the tester clock input and the ATE controls the clock select input so that either the ATE clock or the system clock can be used to clock the system. The test point is a terminal or output where the ATE system can monitor the system clock. Test points can be provided on a system to be tested manually to allow easy connection of test equipment. They may be connectors or special terminals soldered directly into a PCB.

Digital circuit testing is a nontrivial problem, and much research effort has been devoted to devising circuit structures that are capable of being tested quickly. A very important 1990s development in the testing of digital integrated circuits and boards is boundary scan, which allows test access to integrated circuits on a printed circuit board via a serial test bus and also testing of printed circuit board tracks for continuity and short circuits.

Boundary scan is defined in *IEEE Standard Test access port and boundary scan architecture,* IEEE Std. 1149.1 (2001). For an introductory coverage of the topic, see Lewin and Protheroe (1992).

Stress screening

Once a PCB or system has passed a functional test and has been found to work correctly, many manufacturers apply a final series of tests to expose latent defects in the unit that are not readily apparent but will cause early failure. The reason for carrying out these tests was explained in Chapter 9 in the context of the "bathtub" failure rate curve. Latent defects can be detected by operating a product, possibly at an elevated temperature, for a period of days or weeks. This is known as

Stress screening is expensive and is normally applied only to high-value products or to systems where high reliability is essential.

Table 10.4   A selection of simple stress tests for electronic products

Test	Procedure	Detects
Power up/down	Connect the unit to a power source and switch on and off a few times, then test for continued function.	Weak fuses and semiconductors.
Thermal cycle	Run at elevated temperature, then at low or room temperature, and repeat.	Weak and strained joints.
Swept frequency vibration	Mount unit on vibration table and subject to vibration at a single frequency swept through a suitable range repeatedly.	Loose fastenings, poorly mounted components.
Random vibration	Same as previous test but using random vibration.	More effective than previous test because of excitation of resonances.

Burn-in testing can also be carried out by component manufacturers.

burn-in or soak testing. Perhaps a more effective method, however, is to subject the units to be tested to a series of carefully chosen stress tests or stress screens. These tests need not be elaborate or time-consuming, but their effectiveness must be monitored by recording test data for each unit tested and collecting reliability data on the units that fail in service. Table 10.4 presents a summary of some possible stress tests used in electronic product testing. Some of the tests can be applied very easily. Powering the UUT up and then down (switching it on and then off), for example, stresses circuits because of inrush currents and voltage surges that are not present in normal operation and can cause weak components to fail.

Summary

A selection of environmental factors influencing the performance and reliability of electronic systems has been discussed in this chapter. Temperature affects the parameters of many electronic components significantly, and design of precision circuits to operate over a wide temperature range is not easy. Component ageing is accelerated by increased temperature, so that the operating life of systems working at high temperature is reduced. The failure rate of many electronic components may obey the Arrhenius equation, which is a relationship between absolute temperature and the rate of a chemical reaction. One useful effect of the increased failure rate with increasing temperature is the practicality of performing accelerated ageing tests on components by operating them at high temperatures. The effect of water and dust on electronic equipment cannot be dealt with analytically. An empirical classification scheme for the degree of protection provided by an equipment enclosure against water and dust has been described. Vibration, acceleration, and people have been briefly discussed as environmental hazards to electronic systems.

The need for environmental and electromagnetic testing, and the difficulty of testing exhaustively for all possible conditions, has been outlined. The chapter has concluded by introducing the three main types of test applied to production units: in-circuit testing to verify that components have been correctly assembled into a circuit, functional testing to verify that a system operates as intended, and stress testing or screening to detect units with latent manufacturing or component defects.

Problems

10.1 A capacitor has a failure rate of 8×10^{-6} hour^{-1} at 20°C and 2×10^{-5} hour^{-1} at 40°C. If the Arrhenius equation holds for this component, what would be the failure rate at (a) 10°C and (b) 60°C?

10.2 An integrated circuit has a failure rate of 6×10^{-9} hour^{-1} at 20°C. (a) Express the equivalent mean time to failure in years. If the Arrhenius equation applies and the failure process has an activation energy of 0.25 eV, what is the MTTF in years at (b) 0°C and (c) 70°C?

Safety

Objectives

☐ To emphasize the safety responsibilities of design engineers.

☐ To stress the importance of protection against electric shock and to outline the international safety classifications for electrical and electronic equipment.

☐ To discuss briefly some other risks posed by electronic equipment.

☐ To describe some of the design implications of safety standards and legislation.

☐ To describe typical safety tests on mains-operated electronic equipment.

Modern electronic equipment has to be designed for safe operation both in normal use and under fault conditions and also during transit, installation, and maintenance. Safety requirements and recommendations for electronic equipment are contained in a number of technical standards and codes of practice issued by the International Electrotechnical Commission (IEC) and the European Committee for Electrotechnical Standardization (CENELEC).

Unsafe equipment may cause injury or death to the user, resulting in legal action against the equipment manufacturer. For example, in the United Kingdom, Acts of Parliament have been enacted that place legal responsibility for safety on equipment designers and installers, among others. Electronics engineers who fail to take reasonable care over the safety of their design are committing a criminal offence. The most wide-ranging of these British Acts is the Health and Safety at Work, etc. Act 1974, which deals with the health and safety of those at work and includes within its scope any machinery, equipment, or appliance designed for use in a place of work. Specific responsibilities are placed upon designers (and others) to ensure that equipment is safe and without risks to health at all times when it is being set, used, cleaned, or maintained by a person at work, and that adequate information is available about the equipment and any measures necessary to ensure its safe use. In addition, these responsibilities include carrying out research to discover, eliminate, or minimize risks posed by the equipment.

More specific requirements regarding electrical safety are defined by the British Electricity at Work Regulations 1989, which are made under the Health and Safety at Work Act and are an extension of that Act. The design and construction of electronic equipment are subject to the provisions of the Low Voltage Electrical Equipment (Safety) Regulations 1989. These regulations apply to all electrical equipment, including industrial equipment, designed for use between 50 V and 1000 V a.c. or 75 V and 1500 V d.c. (These voltages are known as "low voltage.") The regulations implement a 1973 European Community (EC) directive on electrical equipment designed for use at these voltages, and similar legal implementations exist in other European Union (EU) countries. In

CENELEC stands for Comité Européen de Normalisation Electrotechnique. The organization is based in Brussels and comprises the national electrotechnical committees of the European Union (EU) and Iceland, Norway, and Switzerland.

For a useful general introduction to product safety and liability, see Abbott (1980 and 1997).

This list of responsibilities is not exhaustive.

Trademark of Underwriters
Laboratories, Inc.
(Reproduced with permission.)

Information on the effect of
electric current on the human
body, including the threshold
of perception, is given in DD
IEC/TS 60479-1 (2005).

the United States, the Occupational Safety and Health Administration (OSHA) is the governmental body responsible for (among many other things) electrical safety. Detailed safety requirements for electronic equipment are set out in technical standards, according to the type of equipment and intended use (domestic, commercial, industrial, etc.).

In Europe, mains-operated domestic equipment is approved by a CENELEC member organization. In the United Kingdom, approvals are granted by ASTA BEAB (a body formed in 2004 by the merger of the Association of Short-Circuit Test Authorities and the British Electrotechnical Approvals Board) after tests on a sample of the equipment. The BEAB approval symbol that is marked on approved equipment is shown in the margin. In the United States, Underwriters Laboratories (UL) conducts testing and certification to assess compliance with established standards. The UL certification mark is shown in the margin. Many variants of this mark exist for marking products that meet standards of certain geographical areas such as the USA and Canada. From 1993 the European Union (then the European Community) adopted the Conformité Européenne (CE) mark in the form shown in the margin. This mark denotes conformance to all EU directives applicable to electronic products, including that on electromagnetic compatibility referred to in Chapter 8 and that on equipment designed for use at low voltage. Other directives may, of course, be introduced from time to time.

Electric shock

The most common and dangerous risk posed by electronic equipment is electric shock. Design measures to minimize the risk of electric shock are therefore of fundamental importance. At mains supply frequencies, alternating currents of only 0.5 mA can cause a reaction in healthy people, while 50 mA can be lethal if sustained for more than one second. Figure 11.1 shows a range of alternating currents and time durations, and the regions in which physiological effects occur. Zones AC3 and AC4 in the figure represent dangerous combinations of current and duration. In zone AC-4, effects such as cardiac and breathing arrest, and burns, may occur. In zone AC-3 there may be strong involuntary muscle contractions and difficulty in breathing. In zone AC-2 no harmful electrical physiological effects are likely. Figure 11.2 shows the ranges of direct currents and time durations, and the regions in which physiological effects occur. These differ from the regions and durations for alternating current (direct current is less dangerous). Zones DC1-4 correspond to zones AC1-4 of Figure 11.1 with respect to physiological effects, and therefore zones DC-3 and DC-4 represent dangerous combinations of current and duration. Even nonlethal shocks can cause injury because of involuntary action such as sudden recoil from the source of shock. Protection against electric shock can consist of measures to limit currents to safe levels, irrespective of voltage. Where current is limited to a safe level, contact with live high-voltage conductors can be quite safe. For example, power-supply designs for cathode ray tubes (CRTs) producing up to 30 kV are safe if the design inherently limits current, typically to 200 to 300 µA.

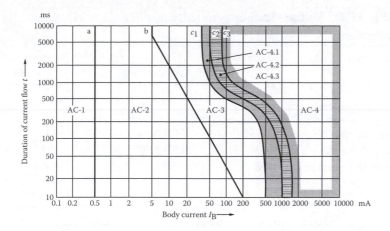

Figure 11.1 Time–current zones of effects of a.c. currents (15 Hz to 100 Hz) on persons for a current path corresponding to left hand to feet. (Reproduced from IEC Publication DD IEC/TS 60479-1:2005. Copyright © 2005 IEC, Geneva, Switzerland. www.iec.ch.)

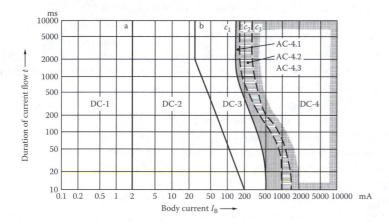

Figure 11.2 Time–current zones of effects of d.c. currents on persons for a longitudinal upward current path. (Reproduced from IEC Publication DD IEC/TS 60479-1:2005. Copyright © 2005 IEC, Geneva, Switzerland. www.iec.ch.)

When discussing safety measures against electric shock, one needs precise definitions of voltage ranges. Table 11.1 lists the three voltage ranges defined by the IEC that are generally applicable to electronic systems. Note carefully that the term *low* does not mean safe: these voltages are only low relative to higher-voltage ranges. Mains supply voltages in common use (120 V, 230 V) are classed as low voltage (LV), but can be lethal. Extra-low voltages (ELV) below 50 V a.c. root mean square (r.m.s.) at mains frequencies, or below 120 V d.c., are normally considered safe under dry working conditions. In wet or highly conducting locations, these voltages are not necessarily safe and other precautions may be needed to prevent danger.

To some extent, the impedance of the human body limits current at low voltages. Current flow from one arm to the other across the chest, however, is especially hazardous because of the danger of electrical interference with the normal action of the heart. One should also be

189

Table 11.1   Voltage ranges generally applicable to electronic equipment

Designation	Between conductors		With respect to earth	
	V a.c. r.m.s.	V d.c.	V a.c. r.m.s.	V d.c.
Low voltage (LV)	<1000	<1500	<600	<600
Extra-low voltage (ELV)	<50	<120	<50	<120
Separated extra-low voltage (SELV)*	<50	<120	<50	<120

* An SELV supply must be isolated from the supply mains and earth by a safety isolating transformer, or be derived from a source, such as a battery, which is independent of a higher-voltage supply.

aware that in some applications, there may be a risk to livestock as well as humans. Some farm animals are much more easily electrocuted than humans because of their low body resistance.

Protective measures against electric shock

There are four possible measures against electric shock that can be applied in the design of electronic equipment. First and most important is inaccessibility of live parts. A live part is defined as any conductor or conductive part intended to be energized in normal use, including a neutral conductor. The need to prevent access to live parts has many design implications, some of which are discussed later in this chapter.

Information on protection against electric shock is contained in the British Institution of Electrical Engineers (IEE) Wiring Regulations, 16th edition, published as BS7671 (2001). Although the Wiring Regulations cover electrical installations rather than electronic systems, many protective measures against electric shock discussed in the regulations are applicable to electronics.

The second protective measure against electric shock is earthing and automatic disconnection of supply in the event of a fault. The concept of an earth or ground is so often taken for granted in electronics that it is worth discussing the idea briefly from a safety viewpoint. When we state or measure a voltage or potential, we refer to an arbitrary reference potential that we designate as 0 V. Quite often the 0 V reference is the protective conductor or "earth" in the mains supply, which is connected to the ground beneath us by a buried rod or plate. To be of any use for safety purposes, the earth connection must be of low impedance, and provided this is so the potential of the earth connection will be effectively unaffected by currents flowing to or from earth because of the large mass of the ground. Any part of an apparatus (such as a metal case) that is securely connected electrically to earth and protected by a suitable automatic disconnection device, such as a fuse or circuit breaker, cannot become live at a dangerous voltage for a significant time. This principle is very widely used.

The third protective measure against electric shock, applicable in some cases, is limitation of voltage to separated extra-low voltage (SELV) as defined in Table 11.1. An SELV supply is limited to less than

The buried rod or plate is known as an earth electrode. In urban areas, this is often located at the substation of the electricity supplier, and each customer supply cable includes a connection wired to the system earth electrodes. In rural areas, the earth electrode may be local to each building supplied.

Very heavy currents can alter the ground potential. For example, if an earth fault occurs on a high-voltage transmission pylon, heavy currents flowing into the ground can raise the ground potential at the base of the pylon. It is possible for a cow or horse to be electrocuted if the animal stands radially close to the pylon.

190

50 V a.c. r.m.s. and must be isolated from the supply mains and earth by a safety isolating transformer or other means. This is safe because these voltages are normally insufficient to cause a hazardous current to flow through the impedance of the human body.

A final protective measure against shock is limitation of current to a safe value, irrespective of voltage. This principle is used in flash testers for testing the insulation of electrical appliances at up to 4 kV. The tester is designed so that current is inherently limited to a safe value.

One final hazard should be pointed out under the heading of electric shock. Reservoir capacitors in electronic equipment can store significant electrical energy that can be hazardous if discharged through the body. Large-value capacitors must therefore be fitted with bleed resistors to dissipate the charge when the equipment is switched off and prevent injury to maintenance personnel. Large-value or high-voltage capacitors can recover a substantial charge after disconnection from a supply if the bleed resistor is not left in place. This is because charge can be absorbed into the dielectric during use. After disconnection from a supply, this charge can be gradually released from the dielectric.

Safety classes for electrical and electronic equipment

IEC publication 61140 (2001) defines four safety classes for equipment with regard to protection against electric shock. These classifications do not indicate quality of protection but rather indicate how protection is achieved. Safety Class 0 equipment relies on basic insulation alone for protection against electric shock, and is not safe unless used in a non-hazardous environment.

The use of Safety Class 0 equipment is not allowed in the United Kingdom.

Safety Class I apparatus is designed to be connected to earth. All accessible conductive parts such as a metal panel, case, or cabinet must be bonded electrically to a protective earth terminal. If the equipment is designed to be connected to a mains supply by a flexible cable, the cable must have a protective earthing conductor and be fitted with a plug with an earthing contact. Either the cable must be nondetachable or the apparatus must have a mains inlet connector with an earthing contact. Live parts must of course be inaccessible, and there must be at least basic insulation throughout the apparatus. Basic insulation is defined as that necessary for proper functioning and basic protection against shock. Safety relies on protective earthing and automatic disconnection of supply to guard against failure of the functional insulation. Safety Class I equipment normally carries the legend: "WARNING: THIS EQUIPMENT MUST BE EARTHED."

Bonded means connected electrically to ensure a common potential. The wire or metal braid used must be of adequate cross-section to carry any fault current.

Safety Class II equipment has no provision for protective earthing and is normally double insulated, all insulated, or equipped with reinforced insulation. Double insulation consists of basic insulation plus independent insulation to ensure protection against shock if the basic insulation fails. Equipment of all-insulated construction must have an outer enclosure constructed entirely from insulating material (with the permitted exception of small metal parts such as screws, which are separated from live parts by a defined thickness of insulation). Reinforced insulation is a single layer of insulation that provides equivalent protection to double insulation. If the outer enclosure of the apparatus is durable and of insulating material, it may be regarded as the second layer of insulation. If the outer enclosure is of metal, double insulation must be used throughout internally. Safety of Class II equipment does not depend on installation conditions such as correct wiring of an earth conductor to a mains supply. Much household equipment is therefore designed as Class II. Examples are hairdryers, power drills, food mixers, and table lamps. Safety Class II equipment is marked with the double-square symbol shown in the margin.

Safety Class III equipment is designed for connection to a SELV supply and does not generate voltages higher than SELV. Protection against shock therefore relies on limitation of voltage. Handlamps for use in ship's bilges (a hazardous wet location) are recommended to be operated from a 25 V SELV supply.

Residual-current circuit breakers

RCCBs are also known as earth leakage circuit breakers, or ELCBs. They are common examples of the wider class of residual current devices, or RCDs. In the United States and Canada, RCDs are known as ground fault circuit interrupters.

For added protection against shock, especially in workshops and electronics development laboratories, special mains circuit breakers are widely used. Residual-current circuit breakers (RCCBs) detect small differences of the order of 10 to 30 mA in the currents flowing in the phase and neutral conductors of a mains supply due to leakage of current to earth (possibly through a human body in contact with a live conductor, as shown in Figure 11.3). The principle of operation of a common type of RCCB is shown in Figure 11.4. The phase and neutral conductors connecting a load to a supply are wound in opposite senses on a toroidal core, and a secondary detector winding is also wound on the core. The phase and neutral conductors form a primary winding, but the load currents in the two conductors are equal and opposite, and generate no net magnetic flux in the core. Any difference current, however, generates an electromotive force (e.m.f.) in the secondary winding that can be used to operate a trip coil to open the circuit breaker contacts.

In some RCCBs, the phase and neutral conductors are simply threaded through the centre of the toroid, each forming a single-turn winding.

RCCBs must be tested regularly and are fitted with a test button for this purpose. Where RCCBs are installed in a workplace, it is a good idea to turn off the mains supply at the end of each day by using the RCCB test button.

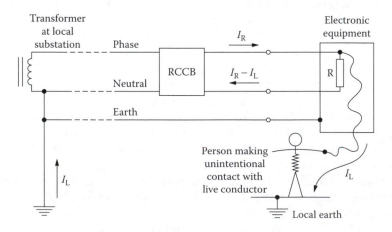

Figure 11.3   Earth leakage.

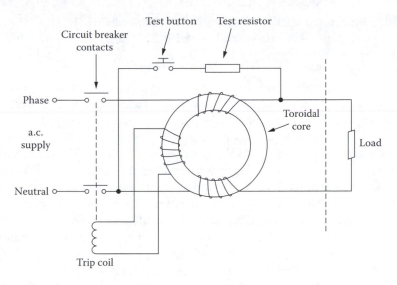

Figure 11.4 Circuit of a typical residual current circuit breaker.

It is important to realize that RCCBs give no protection against contact with phase and neutral conductors simultaneously. Also, the RCCB test button does not verify that the RCCB is operating within its performance specification: it merely checks the electromechanical function of the device.

Electrical safety tests

The British Electricity at Work Regulations 1989 (Statutory Instrument 1989 No. 635) require employers to maintain electrical equipment at work so that it stays safe as far as is reasonably practicable. This requirement will necessitate checking, inspecting, and possibly testing, at intervals to be determined by an employer in the light of the risks. Portable equipment, such as soldering irons, is obviously likely to become damaged in use and should be inspected and tested fairly frequently. Office equipment is less likely to become damaged and may need only a visual check at less frequent intervals. If electronic equipment is to be tested there are several tests that may be carried out. The following assumes that equipment is mains powered, but the need for inspection and testing is not confined to mains-powered apparatus. In the United Kingdom, these tests are often known as Portable Appliance Tests (PAT).

The first test to be done (after a visual inspection for obvious damage) is an earth-bonding test. This test is applicable to Safety Class I apparatus only, and measures the resistance of the earth connection from the equipment case or other accessible metal parts to the mains earth connection (usually the earth pin of the mains plug). Typically, a current of 4 A is used for testing the earth bonding of apparatus with a light supply cable (such as most electronic equipment) and around 25 A for heavier current apparatus. The high current is used to ensure that the earth bonding is capable of carrying a sustained significant current in the event of a fault to give time for a protective device, such as a fuse or circuit breaker, to operate. Typically, a resistance below 100 mΩ would be regarded as satisfactory, so that, for a 25 A test, the voltage developed between the earth contact of the mains plug and the equipment case would be below 2.5 V.

The second likely test is to measure the insulation resistance between the phase and neutral inputs to the appliance (shorted together) and either the earth-bonding connection or the accessible parts of the apparatus for Safety Class I or II apparatus respectively. Test voltages of 500 V d.c., 1.5 kV a.c. r.m.s., or up to 4 kV a.c. r.m.s. may be used. The lower voltage gives an indication of the present state of the insulation and would normally be applied first. A resistance of 10 MΩ or more is desirable. The higher-voltage tests are often known as "flash" tests and may give an early indication of degradation of the insulation. Class II equipment is normally tested at 1.5 kV a.c. r.m.s. As well as testing the insulation, the higher voltages also stress the insulation and may cause early failure if the tests are applied too often. Some equipment is fitted with mains inlet filters, as illustrated in Figure 8.12 for reasons discussed in Chapter 8, and these filters, particularly the capacitors, are unlikely to be rated at 4 kV a.c. This type of equipment should not therefore be subjected to high-voltage insulation tests and should be labelled to this effect.

The final type of test likely to be applied measures the earth leakage current (the difference between the phase and neutral currents when the apparatus is operating). An earth leakage current of more than 500 μA is an indication of a fault, and a much lower figure is normal.

The phase and neutral inputs to the apparatus must be connected during an insulation test, otherwise the test voltage will subject the internal circuitry of the apparatus to more than its normal operating voltage, causing damage.

Other safety hazards

Apart from the risk of death due to electric shock, electronic apparatus can be hazardous in other ways, a few of which are listed in Table 11.2. It should also be realized that quite simple mechanical defects such as sharp edges and inadequate handles could result in injury.

Electric and magnetic fields and electromagnetic waves are, of course, generated very widely by all sorts of electrical and electronic systems, but usually at levels that have in the past been regarded as nonhazardous. Radio-frequency and microwave equipment generating significant power has always been recognized as a potential hazard, especially near to transmitting antennae. In recent years some concern has been expressed about lower-intensity fields and waves, and recommendations or guidance has been published by the International Non-Ionizing Radiation Protection Association, the National Radiological Protection Board (NRPB) in the United Kingdom, and the Institute of Electrical and Electronics Engineers (IEEE) in the United States.

Ionizing radiation in the form of x-rays can be produced in any apparatus that accelerates electrons through voltages greater than 5 kV. CRTs used in older oscilloscopes, televisions, video monitors, and visual display units (VDUs) are the most likely devices to generate x-rays. Recommendations on exposure to ionizing radiation are made by the International Commission on Radiological Protection (ICRP), and an EU directive has been published that is implemented in the United

X-rays are generated when high-energy electrons strike a target material. The mechanism is discussed briefly by Kip (1969).

Table 11.2 A selection of safety hazards in electronic equipment

Hazard	Source	Main risk
Electric and magnetic fields 0–100 kHz	Electrical conductors and circuits.	Adverse physiological effects. Possible long-term effects.
Electromagnetic fields and waves of frequencies between 100 kHz and 300 GHz (including microwaves)	Antennae, electrical conductors and circuits.	Heating, burns. Possible long-term effects.
Ionizing radiation (mainly x-rays)	CRTs with accelerating voltages greater than 5 kV.	Radiation exposure.
Laser radiation	Lasers, laser diodes.	Damage to eyesight.
Toxic gases or fumes	Damaged or overloaded components.	Poisoning.
Sound or ultrasound	Loudspeakers, ultrasonic transducers, switched-mode power supplies.	Hearing damage or loss.
Implosion or explosion	Vacuum tubes, CRTs, overloaded capacitors and batteries.	Injuries caused by flying glass or fragments.
Heat	Hot components, heat sinks.	Burns, fire.

Kingdom by the Ionizing Radiation Regulations 1999 (Statutory Instrument No. 3232).

Hot surfaces on the outside of electronic equipment are an obvious hazard due not only to the risk of burns but also because unexpectedly hot surfaces, handles, or knobs might cause a person to jump away suddenly or drop the equipment. A heatsink may be allowed to reach a higher surface temperature than the rest of an enclosure, whereas knobs, handles, and switches that are intended to be touched are likely to be limited to lower temperatures than general outer surfaces.

Design for safety

So far in this chapter, electronic product safety has been discussed in the abstract. This section discusses some of the design implications of safety legislation and standards.

Inaccessibility of live parts

The need to make live parts inaccessible has many ramifications and often significantly influences the mechanical design of electronic equipment. Any removable cover or panel that exposes live parts must be fastened so that it can only be opened or removed using a tool. Openings for ventilation and so on must be small enough that access by fingers is prevented. A standard test finger is defined in British Standards/European

Norm (BS EN) 60529 (1992) for checking openings where there is doubt as to whether a finger could touch a live part. An even more stringent requirement is that foreign bodies should not touch live parts if introduced into an opening, and BS EN 60529 defines test rods of various sizes for checking this requirement. These requirements could necessitate the provision of some sort of baffle under a ventilation opening or the addition of an insulating cover over live parts.

External plug and socket connections carrying voltages greater than extra-low voltage must be arranged so that any exposed pins or terminals of the connectors are on the dead side of the connection when the connector is separated. When a connector provides a protective earth connection as well as live supply connections, the earth terminal in the connector mates before the live terminals and unmates after them.

Take a look for yourself at a selection of mains connectors such as the British BS (British Standards) 1363 13 A plug or the American NEMA (National Electrical Manufacturers' Association) 5-15 plug, and check this statement.

In many electronic equipment designs, all of the circuitry apart from the primary side of the mains transformer is at extra-low voltage. When the equipment is opened for servicing, no mains conductors should be exposed. Wiring to the transformer or main on/off switch should be fitted with insulating sleeving where it is joined to terminals. If there are mains voltages on a PCB, it is good practice to confine the mains wiring to as small an area of the board as possible and to provide an insulating cover over that part of the board, including the underside, labelled to warn of the danger. The same principles apply where high voltages are generated within a system: the high-voltage parts of the circuit should be partitioned and protected by insulating covers so that servicing personnel are not exposed to danger when the apparatus is under test.

Creepages and clearances

There are specified minimum distances between circuits connected to the mains supply and parts that are accessible. This is to allow for the possibility of a conductive path being established by dust and dirt where there would otherwise be no conduction. A clearance distance is the shortest distance measured in air between two conductive parts, and a creepage distance is the shortest distance measured over the surface of insulation (Figure 11.5). As an example, IEC 61010-1 (2001) gives a creepage distance of 3 mm for a 230 V a.c. r.m.s. voltage on a PCB, or

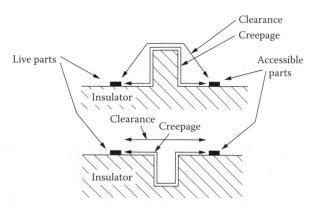

Figure 11.5   Examples of creepage and clearance distances over an insulating barrier and a groove.

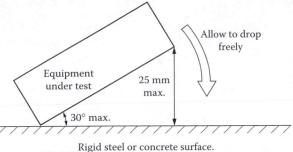

Figure 11.6   Standard drop test.

Rigid steel or concrete surface.
(The test is carried out on all four bottom edges.)

4.6 mm if not on a PCB. The clearance distance at the same voltage is also 4.6 mm.

Mechanical strength

All of the electrical safety features, such as inaccessibility of live parts, must be proof against reasonable mechanical abuse. A drop test, as shown in Figure 11.6, is one test that equipment might be required to withstand without breakage of insulation, cracking of the enclosure, or loosening of covers. There may also be specified vibration and impact tests, and requirements for mechanical integrity to be maintained while the equipment is exposed to heat.

Markings

To comply with safety standards, electronic apparatus must be marked with certain information including the nature of the electrical supply (a.c. or d.c.), the supply frequency if a.c., the rated voltage and the power consumption if more than 25 VA, the off position of the on/off switch if any, and the meanings of any indicator or warning lamps. The markings must be clear, suitably placed, and indelible.

Protection against fire

When a fault occurs in an electronic system, heavy currents can flow, causing overheating of wiring and components. Faults can be due to component failure, mechanical damage, short circuits caused by foreign bodies falling into ventilation apertures, or deliberate abuse. Protection against fire caused by overloads is obtained by automatic operation of devices to interrupt the flow of excessive current.

The cheapest protective device is a fuse: a wire link or metal strip designed to melt rapidly above a certain current and so break the circuit. Fuses are not foolproof, and there is an incidence of bad practice by unskilled people replacing them with fuses of higher rating or even with improvised substitutes such as bent paper clips and nails. For electronic applications, there are two main types of fuse. Quick-acting fuses consist of plain fuse wire or strip inside a glass or ceramic cartridge, and they are designed to rupture quickly once the rated current has been

The replaceable part of a fuse assembly is known as a fuse link.

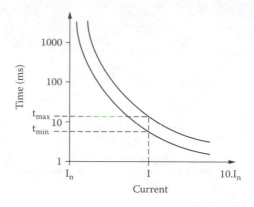

Figure 11.7   A typical fuse time–current characteristic. I_n is the rated current.

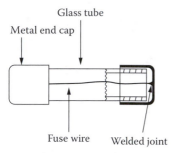

Typical construction of a quick-acting (LRC) fuse link.

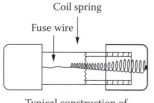

Typical construction of a time-lag (LRC) fuse link.

exceeded. Time-lag or surge-proof fuses are able to carry overload or surge currents of a few times their rated current for several hundred milliseconds. A common application for this type of fuse is in power supplies, where there can be a large inrush current just after switching on as the reservoir capacitors charge up.

The rupturing capacity of a fuse is a very important performance parameter. Quick-acting fuses are available as high-rupturing capacity (HRC) types with a rated rupturing capacity of 1.5 kA, or as low-rupturing capacity (LRC) types with a rated rupturing capacity of 35 A or 10 times the rated current, whichever is greater. Time-lag fuses are of the LRC type. If an LRC fuse operates to disconnect a heavy current greater than its rupturing capacity, the fuse wire may vaporize rather than melt and form a conductive path between the end terminals of the fuse link, either as an arc or along the walls of the fuse link, and the current may not be interrupted. HRC fuses incorporate additional design features, such as a silica filling, to ensure that the arc resulting from operation by a heavy current is quenched and the current is safely interrupted.

Where the additional cost can be justified, perhaps on larger equipment, circuit breakers are to be preferred to fuses, as they cannot be tampered with by unskilled users. They have the additional advantages that they can be reset easily once the fault condition has been cleared and they can break more than one pole of the supply. The performance characteristics of fuses and circuit breakers are shown on time–current characteristics of the form illustrated by Figure 11.7. Two curves on the characteristic show the maximum and minimum times for the device to operate as a function of fault current.

When a protective earth conductor is present, it must never be disconnected by the action of any overload protection device.

Summary

Responsibility for the safety of an electronic system rests primarily with the equipment's designer. Neglect of product safety is a criminal offence in many countries.

The most important danger posed by electronic equipment is electric shock, and protective measures against shock are most important. These include inaccessibility of live parts, protective earthing and automatic disconnection of supply, and limitation of voltage and current. RCCBs can provide additional protection against shock in laboratories and workshops.

Safety Class I equipment achieves protection against shock by having accessible conductive parts connected to a protective earth and relies on automatic disconnection of supply in the event of a fault. Safety Class II equipment does not depend on protective earthing and is either double insulated, all insulated, or equipped with reinforced insulation. Safety Class III equipment operates from an SELV supply.

Safety testing of electronic equipment includes tests of earth-bonding integrity (for Class I equipment), insulation resistance, and earth leakage current.

Other safety hazards posed by electronic equipment include ionizing, microwave and laser radiation, sonic and ultrasonic pressure, heat, and fire. Fuses and/or circuit breakers protect against fire caused by electrical overload.

Some of the design consequences of safety requirements have been discussed, including the need to restrict openings and holes to prevent access to live parts, and the need for mechanical integrity to ensure continued safety.

References

Books

Abbott, H.: *Safe enough to sell?* The Design Council (1980).

Abbott, H.: *Safe by design*; The Design Council (1997).

Anderson, J. C., Leaver, K. D., Alexander, J. M., and Rawlings, R. D.: *Materials science*; Stanley Thornes (4th edition, 1990).

Bannister, B. R., and Whitehead, D. G.: *Instrumentation: transducers and interfacing*; Chapman and Hall (2nd edition, 1991).

Bradley, D. A.: *Power electronics*; Chapman and Hall (2nd edition, 1995).

Carter, R. G.: *Electromagnetism for electronic engineering*; Chapman and Hall (2nd edition, 1992).

Chatterton, P. A., and Houlden, M. A.: *EMC: electromagnetic theory to practical design*; Wiley (1992).

Cluley, J. C.: *Electronic equipment reliability*; Macmillan (2nd edition, 1981).

Compton, A. J.: *Basic electromagnetism and its applications*; Chapman and Hall (1990).

Edwards, P. R.: *Manufacturing technology in the electronics industry*; Chapman and Hall (1991).

Kip, A. F.: *Fundamentals of electricity and magnetism*; McGraw-Hill (2nd edition, 1969).

Kraus, J. D.: *Electromagnetics*; McGraw-Hill (4th edition, 1992).

Lewin, D., and Protheroe, D.: *Design of logic systems*; Chapman and Hall (2nd edition, 1992).

Long, C.: *Essential heat transfer*; Longman (1999).

Meade, M. L., and Dillon, C. R.: *Signals and systems: models and behaviour*; Chapman and Hall (2nd edition, 1991).

Morant, M. J.: *Integrated circuit design and technology*; Chapman and Hall (1990).

O'Connor, P. D. T.: *Practical reliability engineering*; Wiley (4th edition, 2002).

Ott, H. W.: *Noise reduction techniques in electronic systems*; Wiley (2nd edition, 1988).

Ritchie, G. J.: *Transistor circuit techniques*; Stanley Thornes (3rd edition, 1998).

Scarlett, J. A.: *An introduction to printed circuit board technology;* Electrochemical Publications (1984).

Senturia, S. D., and Wedlock, B. D.: *Electronic circuits and applications*; Krieger (1993).

Sparkes, J. J.: *Semiconductor devices: how they work*; Chapman and Hall (2nd edition, 1994).

Szymanski, J. E.: *Basic mathematics for electronics engineers: models and applications*; Chapman and Hall (1989).

Till, W. C., and Luxon, J. T.: *Integrated circuits: materials, devices and fabrication*; Prentice-Hall (1982).

Vincent, C. A., and Scosati, B.: *Modern batteries*; Butterworth-Heinemann (2nd edition, 2000).

Wilkins, B. R.: *Testing digital circuits*; Chapman and Hall (1990).

Williams, T.: *EMC for product designers*; Butterworth-Heinemann (3rd edition, 2001).

Wong, H. Y.: *Heat transfer for engineers*; Longman (1977).

Technical standards

International and national standards may be located by searching the online catalogues of the relevant bodies. For convenience, a short list has been given below:

American National Standards Institute (USA)	www.ansi.org
British Standards Institution (UK)	www.bsi-global.com
European Committee for Electrotechnical Standardization (CENELEC)	www.cenelec.org
Federal Communications Commission (USA)	www.fcc.gov
International Electrotechnical Commission	www.iec.ch

Answers to problems

2.1 3.4 mm^2

2.2 4.9 mm

3.1 6 m^2

3.2 About 30 million

4.1 About 200 mW

4.2 18,000 F

4.3 High temperatures accelerate self-discharge.

4.4 a. 2.5 mF
    b. 5.1 A (no change)

4.5 a. $C/8$
    b. $3C/32$

4.6 a. 62%
    b. 5%

4.7 a. 70%
    b. 35%

5.1 ±0.1%

5.2 $1/\pi$ pF, or about 0.3 pF

5.3 13 mV r.m.s.

6.1 a. $0.76\ \Omega$
    b. −1%, +3%

6.2 a. $240\ \mathrm{k\Omega}$
    b. $20.4\ \mathrm{k\Omega}$

7.1 About $1°\mathrm{C\ W^{-1}}$

7.2 a. 1.5 W
    b. 83°C

7.3 $21\ \mathrm{m^3\ hour^{-1}}$

8.1 Show that this model is equivalent to equation (8.5) in terms of the voltage at the termination.

8.2 a. About 1.5 GHz
    b. −48, −38, −34, −31, and −29 dB respectively for the 1st, 3rd, 5th, 7th, and 9th harmonics (use $20\log_{10}[|v_i|/V]$)

8.3 a. $R_1 = 200\ \Omega$, $R_2 = 300\ \Omega$
    b. 20×10^6 items per second

8.4 a. 200 mA
    b. 12 nF
    c. No account has been taken of the time taken for charge to leave the power-supply reservoir capacitor.

9.1 a. 99.76%
    b. About 1100 hours

9.2 a. 79,000 hours
    b. 0.88

9.3 0.65

9.4 3.3 years

10.1 a. 4.8×10^{-6}
    b. 4.5×10^{-5}

10.2 a. 19,000
    b. 39,000
    c. 4500

Index

A

Absolute maximum ratings 80
A.C. Mains 54
Accelerated life tests 174
Acceptors 33
Accessibility
    of live parts 190
    for maintenance 167
ADC, *see* Analogue-to-digital converter
ADSL, *see* Assymetrical digital
       subscriber line (ADSL)
Ageing 2, 80, 155
Alkaline manganese cell 57
All-insulated 191
Aluminium
    anodized 176
    electrolytic capacitors 90
American National Standards Institute
      (ANSI) 202
Ampere-hour (Ah) 56
Analogue systems 144
Analogue-to-digital converter 98, 108
    dual-slope integrator 99, 137
    single-slope 77, 78
Annealing 41
Anodized aluminium 176
ANSI, *see* American National Standards
      Institute
Antenna 131, 134
Application-specific ICs 46
Arrhenius equation 173-4
Arsenic 33
Artwork 19
Assymetrical digital subscriber line
      (ADSL) 14
Association of Short-Circuit Test
      Authorities (ASTA)–British
      Electrotechnical Approvals Board
      (BEAB) 188
Attenuation
    coefficient 130
    transmission line 14
Automatic test equipment (ATE)181
Availability 158
"Avometer" 102

B

Back driving 180
Backplane 9, 18, 30
Bandwidth 104
    limitation 137, 145
Barium titanate 88
Basic insulation 191
"Bathtub" curve 155
Batteries 53, 55
    primary 31, 55, 57
    secondary 55, 58
BEAB, *see* British Electrotechnical
      Approvals Board
Bed of nails 181
Bleed resistor 191
Blocking diode 60
Bonding
    covalent 33
    earth, test for 193
    electrical 191
    eutectic 42
    ICs 42
    thermocompression 42
Boron 33
Boundary scan 183
Bridge, solder 25
Bridge rectifier 67
British Electrotechnical Approvals
      Board (BEAB) 188
British Standards Institution 202
Buried layer 37
Burn-in testing 156, 184
Bus, IEEE-488 181
Button cell 58

C

"C"-rate 56
Cable 11, 12
    coaxial 13
    ribbon 13, 14, 17
    screened 13
    twin feeder 13
    twisted-pair 13

CAD (Computer-aided design)
  ICs 48
  PCBs 19, 23
Cadmium 55, 58
  sulphide 60
Calibration 166
Capacitance 81
  parasitic 124
  thermal 117
Capacitors 87
  ceramic 89
  dielectric absorption 88
  electrolytic 90
  feedthrough 141
  leadthrough 141
  mica 89
  nonelectrolytic 89
  polymer 89
  reservoir 61, 65, 68, 90, 191
  safety, large values 88, 191
Capacity (of a battery) 56
Card cage 18
Cathode-ray tube (CRT) 105
CE (Conformité Européenne) mark 133, 188
Cell
  alkaline 57
  button 58
  electrochemical 55
  fuel 55
  lead-acid 58
  Leclanché 57
  lithium 58
  nickel-cadmium 58
  photovoltaic 55, 60
  zinc–carbon 57
  zinc–chloride 57
  zinc–mercuric oxide 58
  zinc–silver oxide 58
CENELEC (European Committee for Electrotechnical Standards) 187
Ceramic capacitor 89
Ceramics, high-K 88
Cermet 85
Characteristic impedance 14, 17, 130
Chopping
  in oscilloscopes 106
  in switching regulators 73
Circuit breaker 198
  residual current 192
Cleanliness (in IC fabrication) 35
Clearance distance 196
CMOS (complementary metal-oxide semiconductor) logic 33, 147, 172

Coaxial cable 13
Cockcroft-Walton multiplier 61
Common-mode
  failure 165
  interference 140
Complete failure 154
Complexity, ICs 31
Component
  mounting 20
  placement 23
  ratings 80
Computer-aided design, *see* CAD
Conduction
  electromagnetic 134
  thermal 112
Conformité Européenne, *see* CE mark
Connectors 15
  filtered 141
  ribbon cable 17
  wiring 15
Constantan 85
Convection, thermal 112
Cooling, forced convection 119
Cooling fans 119
Core
  in IC design 48
  transformer 66
Corrective maintenance 166
Corrosion 176
Coupling
  capacitive 125, 126
  inductive 128
  in oscilloscopes 107
  power supply 139
Covalent bonding 33
Creepage distance 196
Crimping 11, 18
Crosstalk 10, 123
Crowbar circuit 64
CRT, *see* Cathode-ray tube
Crystal orientation 37
Current 96
  limiting 64
Custom ICs 46
Czochralski process 36

D

Damage to semiconductors
  electrostatic 45
  thermal 46
Data sheets 77, 113, 115
D.C.–D.C. converter 61
Declaration of Conformity (USA) 133

Decoupling 87
    in logic circuits 148
Delay, time 97
Derating 156, 174
    curve 116
Design
    for function 153
    for maintainability 167
    PCBs 25
    for production 29
    rules 41
    for safety 195
    for testability 181
Dicing, ICs 42
Dielectric
    absorption 88
    constant 87
    high-K 89
    strength 88
Differential mode interference 140
Differential thermal expansion 45
Diffusion 40
Digital
    oscilloscope 107
    PCBs 146
    systems 146
Digital subscriber line, *see* DSL
Diode 33
    blocking 60
    power 67
    Schottky 42, 60
    Zener 69
Direct write-on-wafer 40
Directives, European Union
    EMC 133, 179, 141
    RoHS 55
Discharge, electrostatic 45
Discharge characteristic 56
Discrete semiconductors 31
Discrete wiring 11, 26
Dissipation factor 83
Distributed parameter systems 128,
        147
Diversity 165
Donors 33
Dopants 33
Double insulation 191-2
Double-sided PCBs, good engineering
        practice 149
Drift 107
    failure 154
Drilling, numerically-controlled 22
Drop test 178, 197
DSL 14
Dual-in-line (DIL) package 43

Dual-slope integrator ADC 99, 137
Dust 176
Dynamic impedance
    of a power supply 63
    of a transmission line 130

E

"E" ranges 79
Earth bonding test 193
Earth leakage circuit breaker 192
Earth plane, *see* Ground Plane
Earthing 190
Eddy currents 65, 92, 111
Electric fields 13, 124, 131, 195
Electric shock 188
    protective measures 190
Electrical interference 55, 66, 133
Electrical safety tests 193
Electricity at Work Regulations 187,
        193
Electrochemical cell 55
Electrolytic capacitors 90
Electromagnetic compatibility 133,
        147
Electromagnetic fields 13, 127
    impedance 135
    time-varying 141
Electromagnetic interference 13, 123,
        133, 140
    common-mode 140
    differential mode 140
    ferrite beads 92
    high frequency 125, 140
    in-band 145
    mechanisms 134
    transient 142
Electromagnetic testing 179
Electromechanical meters 100
Electron-beam lithography 40
Electron gun 105
Electrostatic damage 45
Electrostatic discharge (ESD) 45, 46
Electrostatic screen 66, 126, 141
EMC (EU Electromagnetic
        Compatibility) Directive 133,
        179
Energy 53
    losses, *see* Losses in
    sources 54
    storage, in inductor/capacitor 73
Environmental factors 171
Environmental testing 178
Epitaxy 37

Epoxy resin laminate 21
Equipment wire 11
Equivalent circuit
    capacitor 83
    resistor 82
Equivalent series resistance 83
ESD, *see* Electrostatic discharge
Etching
    in IC fabrication 40
    of PCBs 22
Ethernet 14
European Committee for
        Electrotechnical Standards, *see*
        CENELEC
Eutectic alloy, solder 7
Eutectic bonding 42
Excess noise 86
Extra-low voltage 189, 196
Extrinsic silicon 33

F

Fabrication, of ICs 35
Fail safe 164
Failure 2, 154
    common-mode 165
    freak 155
    mechanisms 154
Failure rate 155
    constant 160
    derivation of 159
Fall time 98
Fans, cooling 119
Farad 87
Faraday cage 127, 141
Fault diagnosis 166
Faults 154
Federal Communications Commission
        (USA) 133
Ferric chloride 22
Ferrite 65, 92, 141
    beads 92, 146
Ferroelectrics 88
Ferromagnetics 92
Field, *see* Electromagnetic fields
Field programmable gate array (FPGA)
        50
Field trials 171, 178
Figure of merit 101
Filter, mains inlet 140
Filtering 140, 146
Fire, protection against 197
Flash testing 191, 194
Flashover 96, 175

Flat packs 45
Flex 13
Float zone process 36
Floating ground 138
Flux
    leakage 66
    linkage 128
    magnetic 127
    solder 8
Foldback limiting 63
Forced cooling 119
Foreign bodies 176
FPGA, *see* Field programmable gate
        array
Freak failure 155
Frequency
    measurement of 97, 101
    receiver 103
    standard 103
    of supply 54
    translation 145
F.s.d., *see* Full-scale deflection
Fuel cell 55
Full-custom IC 47
Full-scale deflection (f.s.d.) 100
Full-wave rectifier 65, 67
Functional testing 181
Furnaces, diffusion and oxidation 38,
        41
Fuses 64, 66, 184, 197

G

Gallium arsenide 60
Gasket, radio-frequency 142
Gate array 49
    field programmable 50
Germanium semiconductors 46
Grid, electrical 54
Ground
    currents 138
    loop 138
    plane 126, 139
    plane PCB 146
Grounding 137
Guard ring 144
Guarding 180
Guided waves 130, 140

H

Handling, semiconductors 45
Harmonics 55

Hazards 194
   See also Electric Shock
Health and Safety at Work, etc. Act 187
Heat 2, 111
   ladder 121
   transfer 112
Heat sink 114
Heat sink calculations
   steady-state 116
   transient 117
High-K ceramics 88
High reliability systems 162
Holes
   plated-through, *see* Through-hole
      plating
   via 23
Humans (as environmental hazards)
   178
Humidity 174
   relative 176
Hybrid microcircuits 42, 85

I

ICs, *see* Integrated Circuits
Ideal properties 2
IEC, *see* International Electrotechnical
   Commission
IEEE (Institute of Electrical and
   Elecronics Engineers)-488 bus
   181
Impedance
   a.c. (of a power supply) 63
   characteristic 14, 17, 130
   of electromagnetic field 135
   of free space 135
   thermal 118
In-band interference 145
In-circuit testing 180
Inaccessibility (of live parts) 195
Inductance 81
   parasitic 127
Induction fields 135
Inductors 91
Infant mortality 155
Infrared reflow 24
Instability 123
Insulation 191
   resistance 17, 194
Insulation displacement 10, 17
Integrated circuits (ICs)
   application specific 46
   CAD 48
   custom 46
   fabrication 35
   full-custom 47
   gate array 49
   standard cell 48
Interconnection 1, 5
Interference, *see* Electromagnetic
   interference
Intermodulation 145
International Electrotechnical
   Commission (IEC) 187, 202
Intrinsic impedance of free space 135
Intrinsic silicon 33
Inverters 61
Ion implantation 40
Ionizing radiation 194
Isolation 66
   from supply mains 65, 191

J

Johnson noise 86
Jointing 1, 5
   crimping 11
   soldering 6
   welding 11
   wire-wrap 9

L

Laminate 21
Large-scale integration (LSI) 32, 45
Laser radiation 194
Latent defects 183
LCCC, *see* Leadless ceramic chip carrier
   package
Lead 6, 55
Lead-acid cell 58
Lead-free solder 6, 8
Leadless ceramic chip carrier (LCCC)
   package 43
Leakage resistance 144, 176
   of a capacitor 77, 88, 144
Leclanché cell 57
Legends 22
Lifetime distribution function, F(t) 159
Light, speed of 129, 147
Linear regulator 70
Linearity 77
Linewidth (in IC fabrication) 40
Lithium cell 58
Lithography 38
   electron beam 40
Live parts 190, 195

Load regulation 62
Loss angle 83, 88
Losses in
    capacitors 111
    dielectrics 83, 88
    filters 140, 141
    inductors 92
    transformers 65
Low voltage (LV) 190
LSI , *see* Large-scale integration
Lumped parameter circuits 81, 130
LV, *see* Low voltage

M

Magnetic field 13, 195
Magnetic flux 127
Magnetic shield 128, 141
Mains supplies, a.c. 54
Maintenance 165
    corrective 166
    preventive 156, 166
Manganin 85
Markings, requirements for 197
Mask (photomask) 39
Mask programming 50
Mass soldering 24, 46
Mean time between failures (MTBF)
    73, 157
Mean time to repair (MTTR) 157
Medium scale integration (MSI) 32
Mercury 8, 55
    See also Zinc–mercuric oxide cell
Metallization 41
Mica capacitors 89
Mica washers 115
Microphony 123
Milling, computer controlled 18
Monotonic drift 154
Moore's law 31
Motherboard 18
Mounting
    components 20
    power transistor 115
Moving coil meter 100
Moving iron meter 101
MSI, *see* Medium scale integration
MTBF, *see* Mean time between failures
MTTR, *see* Mean time to repair
Multilayer PCBs 19, 22
    good engineering practice 149
Multimeters 98, 101
Mumetal 128
Mutual inductance 127

N

"n"-type silicon 33
NC drilling 22
Near field region 135
Nichrome alloy 85
Nickel–cadmium (NiCd) cell 58
Nickel metal hydride (NiMH) cell 59
Noise 85, 106
Nominal value 79
Numerically-controlled (NC) drilling 22

O

Occupational Safety and Health
    Administration (OSHA) (U.S.)
    188
One-shot waveform 108
Open-circuit voltage (of a battery) 56
Operating temperature 111
Oscilloscope 104
OSHA, *see* Occupational Safety and
    Health Administration
Overload protection 63
Overvoltage protection 64
Oxidation, thermal 38
Oxide growth 38

P

"p"-type silicon 33
Packaging
    protection against electrostatic
        discharge 46
    semiconductors 43
Parallel systems (reliability) 163
Parasitic
    behaviour of components 80
    capacitance 124
    circuit elements 123
    effects 2
    inductance 127
    load 108
    reactance 124
    resistance 123
Partial failure 154
Passivation 38
PCB, *see* Printed circuit board
Peak inverse voltage 68
Period 97
Permeability 91, 130, 135
Permittivity 87, 130, 135
PGA, *see* Pin grid array

Phase diagram 7, 8
Phase shift 97
Phosphor 105
Phosphorus 33
Photolithography 39
Photomask 39
Photoplotting 19
Photoresist 21, 39
Photovoltaic cell 55, 60
Pickup 13, 123
    ferrite beads 92
Pin grid array (PGA) 43, 45
PLA, *see* Programmable logic array
Plane-wave electromagnetic field 134
Plastic-leaded chip carrier (PLCC)
        package 43
Plated-through holes, *see* Through-hole
        plating
PLCC, *see* Plastic-leaded chip carrier
        package
Polarization 87
Polycrystalline silicon 36
Polymer film capacitors 89
Polymer films 88
Polytetrafluoroethylene (PTFE), 11
Polyvinyl chloride (PVC), 11
Potentiometers 86
Power 53
    density 121
    diodes 67
    factor 83
    transistor, heat sink mounting 115
Power supplies 60
    uninterruptible 61
Power supply coupling 139
Pre-preg 22
Preferred values 79
Presets 86
Preventive maintenance 166
Primary batteries 31, 55, 57
Printed circuit board (PCB) 1, 5, 18
    assembly 24
    CAD 19, 23
    design 23
    etching 22
    flexible 20
    good engineering practice 145
    manufacture 21
    materials 21
    multilayer 20, 22
    repair 25
    rigid 19
PROM, *see* Programmable read-only
        memory
Process (in IC fabrication) 35

Production testing 180
Production yield 42
Programmable logic array (PLA) 50
Programmable read-only memory
        (PROM) 50
Protection
    against electric shock 190
    against electrostatic discharge 46
    against fire 197
    overload 63
    overvoltage 64
PTFE, *see* Polytetrafluoroethylene
PVC, *see* Polyvinyl chloride

Q

Q factor 83, 89
Quality control 45, 156

R

Radiation
    electromagnetic 134
    ionizing 194
    laser 194
    thermal 113
    X-ray 194
Radio-absorbing material (RAM) 179
Rail, supply 53
RAM, *see* Radio-absorbing material
Ratings
    components 80
    connectors 17
    rectifiers 67
    transformers 66
    wire 11
RCD, *see* Residual current device
Read-only memory (ROM) 50
Recommended ratings 80
Rectification 67
    precision 98
Redundancy 163
Reed relays 92
Reference voltage 69, 96
Reflow soldering 24
Regulation (in a power supply) 62
Regulators
    IC voltage 72
    linear 70
    switching 73
Reinforced insulation 191
Relative humidity 176
Relays 92

Reliability 2, 111, 153, 158
    effect of temperature 173
    function, R(t) 159
    IC packages 43
    ICs 31
    voltage regulators 73
Reliability models
    parallel 163
    series 162
Repair 166
    PCBs 25
Reservoir capacitors 61, 65, 68, 90, 191
    hazard from 88, 191
Residual current circuit breaker 192
Residual current device (RCD) 46, 192
Resist 21, 39
    solder 22
Resistance 81
    insulation 17
    thermal 113
Resistivity, sheet 85
Resistor networks, thick film 85
Resistors 83
    carbon film 85
    metal film 81, 85
    precision wire-wound 85
Restriction of Hazardous Substances,
    EU Directive 55
Rework 25
Ribbon cable 13, 14
    connectors 17
Ripple 54, 68
Rise time 98
RJ-11/RJ-45 jacks 15, 17
RoHS, see Restriction of Hazardous
    Substances, EU Directive
ROM, see Read-only memory
Root mean square 54, 96
Routing 23
Rupturing capacity (of a fuse) 198

S

Safety 2, 187
    classes (for electronic equipment) 191
    See also Electric shock
    hazards 194
    isolation, transformers 65
    testing 193
Sample and hold circuit 143
Sapphire 37
Scaling (of IC devices) 32
Schmitt trigger 102
Schottky diode 42, 60

SCR 64
Screen, electrostatic 66, 126, 128, 141
Screen printing 23
Screened cable 13
Screw terminals 17
Second sourcing 50
Secondary batteries 58
Secondary failure 155
Selenium 60
Self-discharge 56
Self-inductance 124, 127
    of power supply loop 148
SELV, see Separated extra low voltage
Semiconductor packaging 43
Semiconductors, handling 45
Sensitivity 100
Separated extra low voltage (SELV) 190
Series systems (reliability) 162
Sheet resistivity 85
Shield 128
Shielding 141
    effectiveness 142
Shock
    see Electric shock
    thermal 45
Shot noise 86
Shunt resistor 100
Signal-to-noise ratio 86
Silica 36
Silicon 33
    amorphous 60
    dioxide 36, 38
    nitride 38
    polycrystalline 36
Silicone grease 115
Single-slope ADC 77
Sintering 42
Slope resistance (of a Zener diode) 69
Small outline (SO) package 43, 44
Small-scale integration (SSI) 32
Snubber network 142
SO package, see Small outline package
Solder 1
    bridges 22
    eutectic alloy 7
    lead-free 7
    phase diagram 7, 8
    removal 25
    resist 22
    wire 8
Solderability 7, 8
Soldering 6
    mass 24, 46
    reflow 24
    wave 24

Spade terminal 18
Speed of light 129, 147
S.r.d.p., *see* Synthetic-resin-bonded
  paper
SSI, *see* Small-scale integration
Stability 80
  resistors 84
Stabilization 62
Standard-cell ICs 48
Stress 80
  mechanical 176
  screening 156, 183
Substrate 33, 37
Sulphuric acid 58
Supply rails 53
Suppression 133, 142
Surface mounting 20, 45
Susceptibility, EMI 136
Switched-mode power supplies 73, 91
Switching regulator 73
Synthetic-resin-bonded paper (s.r.b.p.)
  21

T

tan δ 83
Tantalum electrolytic capacitors 90
Tantalum pentoxide 90
Temperature 171
Temperature coefficient
  of breakdown voltage 70, 172
  of component parameters 79, 172
  of expansion 173
  power supply 63
  of resistance 84
  of Zener diode 72
Temperature dependence of component
  parameters 79, 172
Terminals
  eyelet 18
  screw 17
  spade 18
Termination (of transmission line) 131
Test chambers 178, 179
Test equipment 95
Test finger 195
Test points 183, 167
Testability 181
Testing 166
  accelerated life 174
  burn-in 156, 184
  drop 178, 197
  for electrical safety 193
  electromagnetic 179

environmental 171, 178
flash 191, 194
functional 181
ICs 42
in-circuit 180
portable appliance 193
production 180
type 178
Thermal
  capacitance 117
  conduction 112
  convection 112
  cycling 80, 112, 173
  damage (to semiconductors) 7, 46
  expansion, differential 45
  grease 115
  impedance 118
  oxidation 38
  radiation 113
  resistance 113
  shock 45
Thermionic valve 31
Thermocompression bonding 42
Thick-film hybrid circuits 85
Through-hole mounting 20, 43
Through-hole plating 19, 22
Thyristor 64
Time-varying electromagnetic fields
  135, 141
Tin 6
Tinning 8
Titanium dioxide 88
TMR, *see* Triplicated modular
  redundancy
Total failure 154
Tolerance
  of components 78
  voltage (of a supply rail) 54
Toxic gases 195
Toxicity, batteries 55
Trade-off 176
Transducers 95
Transformers 65
Transient interference 142
Transistor–transistor logic 147, 172
Transmission line 14, 17, 130
Trickle charging 59
Triggering 106
Trimmer 86
Triplicated modular redundancy (TMR)
  163
Twin feeder 13
Twisted pair 13
Twisted wiring 128
Type testing 178

U

Ultrasonic welding 42
Ultraviolet (UV) light 21
Undercutting 40
Underwriters Laboratories 188
Uninterruptible power supply (UPS)
    61

V

VA, *see* Volt-amperes
Valve, thermionic 31
Vapour-phase reflow 24
Very large-scale integration (VLSI)
Via holes 23
Vibration 176
VLSI, *see* Very large-scale integration
    32
Voltage 96
    drop 12
    doubling 131
    ranges 190
    reference 69, 96
    regulators, IC 72
    standing wave ratio (VSWR)
        17
    tolerance (of a supply rail)
        54
Volt-amperes (VA) 66
VSWR, *see* Voltage, standing wave
    ratio

W

Wafer-scale integration (WSI) 32
Wafers 34
    preparation of 36
Water 174
Watt-hour (Wh) 56
Wave-soldering 24
Wear-out 155
Welding 11
    ultrasonic 42
White noise 86
Windows (in IC fabrication) 33, 40
Wire 11
    gauge 11
Wire-wrap jointing 9
Wiring, discrete 11, 26
Wiring connectors 17
WSI, *see* Wafer-scale integration

X

X-rays 194

Z

Zener diode 69, 72
Zinc–carbon cell 57, 59
Zinc–chloride cell 57, 59
Zinc–mercuric oxide cell 58
Zinc–silver oxide cell 58